Urban Insect Pests: Sustainable Management Strategies

Urban Insect Pests: Sustainable Management Strategies

Editor

M. M. Sawant

scitus
academics

Urban Insect Pests: Sustainable Management Strategies

Edited by **M. M. Sawant**

ISBN: 978-1-68117-103-6

Library of Congress Control Number: 2015935503

SCITUS Academics LLC,
www.scitusacademics.com
Box No. 4766, 616 Corporate Way,
Suite 2, Valley Cottage,
NY 10989

Printed in United States of America on Acid Free Paper ∞

Contents

Preface

Increased urbanization as a result of growing global population refers that sustainable integrated pest management (IPM) strategies for urban pest control are authoritative as these pests are directly linked to the health and welfare of domesticated animals and people. In our fragile urban ecosystem, the risk and safety of exposure to pesticides is disproportionate in view of the high concentration of human population and the sensitive environment. At the same time, regulatory requirements to reduce inadvertent exposures to insecticides in the urban environment are being implemented. The development of insecticide resistance (IRM) strategies in urban ecosystems involves an understanding of the state and mechanisms used by various chemical and nonchemical methods for reducing the pressure of insecticide resilience and insecticide resistance. Pest control is often more affordable through prevention instead of trying to control an established pest population. Moreover, the chance of major economic losses or perhaps irreversible damage is decreased by the prevention of pesticides. Prevention stops various disruptions in connection with control efforts necessary to the establishment of pests.

This book focuses on the widespread molecular and physiologic insecticide resistance approaches of major urban pests. It will also discuss a number of possible methods and tactics for future integrated urban pest control programs. The book focuses on urban insect pest problems to discuss strategies for control that go beyond products. It is not always useful to spray chemicals indoors or in urban environments from the environmental and health point of view so this this book discusses sustainable control and best practice methods for controlling insects like disease vectors, nuisance pests, and structural pests.

Chapter 1

Trends in DDT and Parathyroid Resistance in Anopheles gambiae s.s. Populations from Urban and Agro-industrial Settings in Southern cameroon

Philippe Nwane[1, 2], Josiane Etang[1, 3], Mouhamadou Chouaibou[1, 2], Jean Claude Toto[1], Clément Kerah-Hinzoumbé[4], Rémy Mimpfoundi[2], Herman Parfait Awono-Ambene[1], and Frédéric Simard[5]

[1]Organisation de Coordination pour la lutte contre les Endémies en Afrique Centrale, Yaoundé, Cameroun

[2]Université de Yaoundé I, Yaoundé, Cameroun

[3]Faculty of Medicine and Pharmaceutical Sciences, University of Douala, Cameroun

[4]Programme National de Lutte contre le Paludisme, N'Djamena, Tchad

[5]Institut de Recherche pour le Développement (IRD), UR016, Bobo-Dioulasso, Burkina Faso

ABSTRACT

Background

Pyrethroid insecticides are widely used for insect pest control in Cameroon. In certain insect species, particularly the malaria vector *Anopheles gambiae*, resistance to this class of insecticides is a source of great concern and needs to be monitored in order to sustain the efficacy of vector control operations in the fields. This study highlights trends in DDT and pyrethroid resistance in wild *An. gambiae* populations from South Cameroon.

Methods

Mosquitoes were collected between 2001 and 2007 in four sites in South Cameroon, where insecticides are used for agricultural or personal protection purposes. Insecticide use was documented in each site by interviewing residents. Batches of 2-4 days old adult female mosquitoes reared from larval collections were tested for susceptibility to DDT, permethrin and deltamethrin using standard WHO procedures. Control, dead and survivors mosquitoes from bioassays were identified by PCR-RFLP and characterized for the *kdr* mutations using either the AS-PCR or the HOLA method.

Results

Four chemical insecticide groups were cited in the study sites: organochlorines, organophosphates, carbamates and pyrethroids. These chemicals were used for personal, crop or wood protection. In the four *An. gambiae* populations tested, significant variation in resistance levels, molecular forms composition and *kdr* frequencies were recorded in the time span of the study. Increases in DDT and pyrethroid resistance, as observed in most areas, were generally associated with an increase in the relative frequency of the S molecular form carrying the *kdr* mutations at higher frequencies. In Mangoum, however, where only the S form was present, a significant increase in the frequency of *kdr* alleles between 2003 to 2007 diverged with a decrease of the level of resistance to DDT and pyrethroids. Analyses of the *kdr* frequencies in dead and surviving mosquitoes showed partial correlation between the *kdr* genotypes and resistance phenotypes, suggesting that the *kdr*mechanism may act with certain co-factors to be identified.

Conclusion

These results demonstrate the ongoing spread of *kdr* alleles in *An. gambiae* in Central Africa. The rapid evolution of insecticide resistance in this highly dynamic and genetically polymorphic species remains a challenge for its control.

BACKGROUND

Agriculture and timber production are the main components of the economic activities in most equatorial African countries. These economic activities require intensive use of pesticides including insecticides belonging to the four main chemical groups used in public health: organochlorines, carbamates, organophosphates and pyrethroids. In Cameroon, organochlorines such as DDT have been used extensively for both vector control and agricultural purposes in the southern areas of the country, especially during the 1950s malaria eradication campaign [1-3]. However, these compounds have progressively been replaced by alternative more specific and less toxic chemicals, in part because of the emergence of insecticide resistance in the target species [1,4]. Nowadays, pyrethroids are largely recommended because of their high effectiveness and strong excito-repellent effect on insects, as well as low mammalian toxicity [5-8]. However, the extensive exposure of insect pests to these insecticides has already selected resistance in wild insect populations [9,10]. To date, more than 500 species of insects and mites have been reported to develop resistance to about 300 insecticide compounds [11-13]. Among these species, 56% are crop pests, 39% are arthropods of medical or veterinary importance and 5% are beneficial species[14].

The emergence of insect resistance to insecticides may decrease crop productivity [15] or reduce the effectiveness of insecticide treated nets or indoor residual spraying [13,16]. Resistance management is therefore a major challenge for vector control programmes in countries where vector-borne diseases are endemic and subsistence remains a burden to the communities. Malaria is the most devastating of all vector-borne diseases; it impedes on economic development not only by causing premature death but also through lost/diminished productivity, absenteeism, huge medical cost, and negative impact on fertility, population growth, and country's savings and investments [17].

Several insecticide resistance mechanisms have been reported in different classes of insects of medical and economic importance, particularly among major malaria vectors belonging to the*Anopheles gambiae* complex [18-21]. These mechanisms include enhanced detoxification of insecticides through increased enzymatic activities of esterases, gluthatione S-transferases and cytochrome P_{450} monooxygenases, mainly due to their overproduction as a result of gene amplification [22,23] and/or gene regulation [24-26]. Point mutations at the target sites of insecticides, decreasing the affinity of the insecticide to its receptor, constitute the second major and most widespread mechanism by which insects are able to resist insecticides [27]. Two mutations at amino acid position 1014 of the voltage-gated sodium channel, changing either a Leucine residue to a Phenylalanine (L1014F) [28], or a Leucine to a Serine (L1014S) [29] have been identified in *Anopheles gambiae* and confer knockdown resistance (*kdr*) to DDT and pyrethroid insecticides. On the other hand, organophosphates and carbamates are acetylcholinesterase inhibitors; structural changes in this enzyme in *Drosophila melanogaster,Musca domestica* and *An. gambiae* were reported to cause resistance to these insecticides [30-33].

In Cameroon, cross-sectional surveys of *An. gambiae* s.l. susceptibility to insecticides have been carried out since 1997. Metabolic resistance was suspected in some *An. gambiae* populations in 2003 and confirmed in 2007 [34,35]. More recently, the presence of both *kdr* mutations and their relationship with phenotypic resistance to pyrethroids and DDT were reported in Cameroon[36,37]. Here, we explore trends in insecticide resistance in wild *An. gambiae* populations from Southern Cameroon, through the longitudinal follow-up of insecticide susceptibility levels in four sites where insecticides are used extensively for agricultural or personal protection purposes.

METHODS

Study Sites and Insecticide Usage

The study was conducted in 4 localities (Figure 1) of Southern Cameroon that are characterized by high insecticide usage for agro-industry or personal protection: (1) Ipono district (2°22′N, 9°50′E), an area of intensive forest exploitation and timber storage, with agricultural activities limited to the households' supplies, (2) Bonanloka district (4°03′N, 9°43′E), a highly urbanized area in the centre of Douala, the major economic city in Cameroon, where inhabitants use insecticides against arthropod nuisance, (3) Nkolondom district (3°51′N, 11°30′E), a market gardening area located in the outskirts of Yaoundé, the capital city of Cameroon, and (4) Mangoum district (5°31′N, 10°37′E), a locality with extensive manual and mechanized agricultural settings producing spices, vegetables and cereals. These sites are among the places in Cameroon where agricultural and forest exploitation practices are sustained by pesticide application.

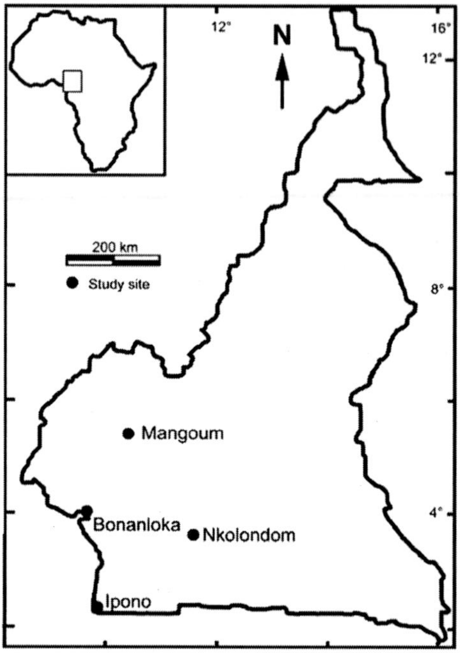

Figure 1: Map of Cameroon showing study sites.

These localities are geographically located in the southern region of the country under the equatorial climate, with average yearly rainfall above 1,500 mm spread out over 4 seasons: 2 dry seasons (December-February and July-August) and 2 rainy seasons (March-June and September-November). However, noticeable variations of these climatic trends are observed in Mangoum located in the western mountain grassland characterized by one dry season between November and February, and one rainy season between March and October [38]. Mosquito larvae were collected during the rainy seasons in 2001, 2003, 2005 and 2007.

During mosquitoes sampling in 2003, a qualitative survey was conducted in each site with the aim to document the most widely used insecticide compounds and agricultural practices. Data concerning insecticide usage in public health, personal protection and agro-industry were collected by filling a standardized questionnaire. Questions asked to local inhabitants, gardeners or owners of timber yards were mainly focused on: (1) insecticides used (trade names/ active ingredients), (2) cultivated crops and their treatment cycles, (3) operational dosages during crop treatments and (4) exploited surface for plant farming or wood storage.

Mosquito Collections and Bioassays

Anopheline larvae were collected from a wide range of breeding sites, representative of the diversity of the mosquito population in each study site. Larvae were collected from puddles, flooded furrows, shallow wells, tire tracks, ponds and marshes. In each location, larval collections were performed in at least 20 breeding sites in which an average of 30 larvae (all instars) per breeding site were collected and reared locally to adults, and fed with a 10% sucrose solution. Upon emergence, mosquitoes were sexed and morphologically identified using morphological identification keys [39,40]. Only females *An. gambiae* s.l. were used for insecticide resistance monitoring. Susceptibility tests were carried out using WHO insecticide susceptibility test-kits and standard protocol for adults [41]. Impregnated filter papers (4% DDT, 0.75% permethrin and 0.05% deltamethrin) were provided by the Vector Control and Research Unit, University Sains Malaysia (Penang, Malaysia). Bioassays were performed at a temperature ranging from 25 to 28°C on 2-4-days old females. For each test, 80-100 mosquitoes separated into 4 batches were exposed to impregnated filter papers, while a batch of 20-25 mosquitoes served as control. The number of knockdown mosquitoes was recorded at 5 min intervals during 1 h exposure and mortality was determined 24 h post-exposure. At each sampling period, bioassays were concomitantly carried out with the Kisumu strain of *An. gambiae* maintained in the Laboratoire de Recherche sur le Paludisme at OCEAC (Yaounde, Cameroon) and used as the reference susceptible strain. Assayed samples were preserved individually on dessiccant (silica gel) and stored at -20°C for further analysis.

Molecular Identification and ***Kdr*** Genotyping

Upon completion of the susceptibility tests and recording of the individual phenotypes, random samples of mosquitoes from bioassays batches at each study period e.g. control, dead and surviving specimens were subjected to DNA extraction [42]. Specimens were identified to species and molecular form by PCR-RFLP [43]. Their genotype at the *kdr* locus was determined using either the Allele-Specific PCR (AS-PCR) [28] in samples collected in 2001 and 2003, or the Hot Oligonucleotide Ligation Assay (HOLA) [44] in samples collected in 2005 and 2007.

Data Analysis

Insecticide susceptibility data were analysed according to WHO criteria [41]: samples were defined as resistant if they showed less than 80% mortality; a mortality rate between 80-98% suggested reduced susceptibility but resistance needs to be confirmed, while mortality rates greater than 98% were indicative of complete susceptibility. The knockdown times for 50 and 95% of tested mosquitoes (KdT_{50} and KdT_{95}) were estimated using a log-time probit model [45]. The KdT_{50}recorded from field-collected mosquitoes were compared with that of the *An. gambiae* Kisumu reference susceptible strain by estimates of KdT_{50} ratios (RR). Chi-square tests were used to compare the prevalence of M and S molecular forms of *An. gambiae* between the different study periods.

RESULTS

Insecticide Usage in Study Sites

More than 30 forms were filled in each study site, except in the Ipono timber yards where only one form was filled because chemical insecticides were not used for another purposes in this study area. Chemicals in use during year 2003 differed from one site to another. Table 1 summarises data collected from surveys conducted in the agricultural settings of Nkolondom and Mangoum. In these sites organophosphates (dimethoate, diazinon, chlorpyrifos-ethyl) and pyrethroids (lambda-cyhalothrin, cypermethrin and deltamethrin) were commonly used for crop protection. In addition, organochlorines (endosulfan and fipronil) and carbamates (carbofuran, methyl-parathion) were also used, especially in Nkolondom. However, large variability among gardeners in application dosages, the crop treatment cycles and the amounts of active ingredients used in these agricultural settings prevented quantitative analysis of the data.

Table 1: Insecticide usage in agricultural settings

Study site	Trade name (concentration)	Active ingredient	Class of insecticide and usage frequency (%)	Cultivated crops
Nkolondom	Bastion (100 g/kg)	Carbofuran	Carbamates (33)	Cabbage, parsley*
	Penncap.M (240 g/l), Sevin (850 g/kg	Methyl-parathion		Celery*, lettuce*,
	Callidim (400 g/l) Cyperdim (200 g/l)	Dimethoate	Organophosphorous (22)	Cabbage,
	Basudine (600 g/l)	Diazinon		Pepper*
	Thionex (500 g/l; 350 g/l), Thiodan (250 g/l 350 g/l)	Endosulfan	Organochlorines (20)	Tomato, spinach*
	Cypercal (12 g/l; 50 g/l; 100 g/l)	Cypermethrin	Pyrethroid (25)	Celery*, tomato, green bean, courgette, basil*
	Décis (12,5 g/l; 25 g/l; 60 g/l)	Deltamethrin		Eggplant
	Karate (2,5%; 5 g/l; 45 g/l)	Lambdacyhalothrin		Tomato, ginger
Mangoum	Thiodan (250 g/l; 350 g/l)	Endosulfan	Organochlorines (8)	Tomato*
	Regent (50 g/l)	Fipronil		Cabbage*
	Callidim (400 g/l), Cyperdim (200 g/l), Dimezyl (400 g/l), Planthoate (400 g/l)	Dimethoate	Organophosphorous (49)	Cabbage*, bean, melon*, watermelon*, tomato* lettuce, nightshade*
	Dursban (600 g/l), Pyriforce (600 g/l)	Chlorpyrifos-ethyl		Corn
	Pilori (15 g/l)	Lambdacyhalothrin	Pyrethroids (43)	Leek*, carrot
	Decis (25 g/l; 60 g/l; 12,5 g/l)	Deltamethrin		Green bean

	Méteor (400 g/l), Cigogne (50 g/l; 200 g/l), Cypercal (12 g/l; 50 g/l; 100 g/l), Cyperplant (100 g/l), Cythrine (25 g/l)	Cypermethrin		Green bean, corn, tomato*, carrot, leek*, nightshade*, potato

*: main cultivated crop

Nwane et al. BMC Infectious Diseases 2009 9:163, doi:10.1186/1471-2334-9-163

In Bonanloka (urban area), bomb sprays and coils containing pyrethroid insecticides with cyfluthrin, deltamethrin, lambda-cyhalothrin as common active ingredients, were mainly used for personal protection. In the timber yard in Ipono, pyrethroids (bifenthrin and cypermethrin) and organochlorines (lindane) were used in insecticide-fungicide mixtures to protect timber against xylophagous insects.

Susceptibility to Insecticides

Throughout the assays, the Kisumu strain of *An. gambiae* displayed mortality rates above 99% for the 3 insecticides tested and no significant variation was observed in knockdown times during the study periods. The kdT_{50} values were approximately 19.0; 9.5 and 8.5 minutes respectively for DDT, deltamethrin and permethrin. The corresponding kdT_{95} values were around 30, 25 and 15 minutes. In control groups of the Kisumu strain as well as that of the wild *An. gambiae* populations (unexposed mosquitoes) mortality rates 24 hours post-exposure were always below 5%. Mortality rates in field mosquito populations are shown in Figure 2 and their respective kdT_{50} and kdT_{95} are given in Table 2. Overall, the four *An. gambiae* populations showed different levels of resistance to the three insecticides from one year to another. In most cases, resistance was associated with an increase of knockdown times compared with the Kisumu susceptible strain, with a kdT_{50} ratio higher than 2 (Table 2).

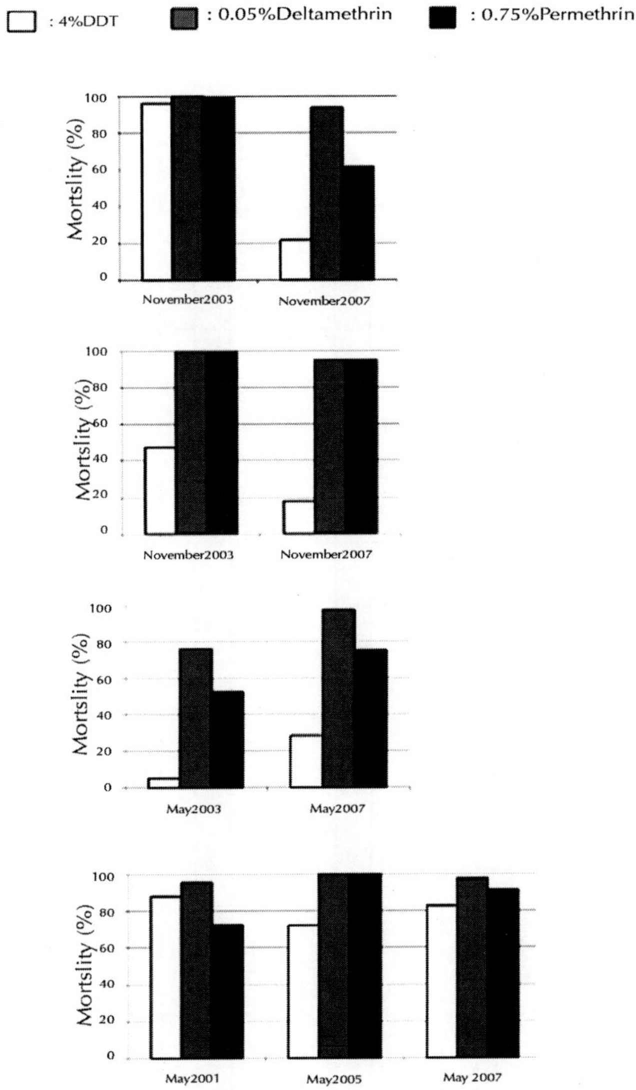

Figure 2: Mortality of *Anopheles gambiae* 24-hours post one-hour exposure to insecticide-impregnated papers at each study period. (A): Ipono, (B): Nkol-ondom, (C): Mangoum, (D): Bonanloka.

Table 2: Knockdown time (kdT) for 50% and 95% of tested *Anopheles gambiae* s.l populations

Site	Insecticide	Period	N	$kdT_{50}[CI_{95}]$ (min)	**$kdT_{95}[CI_{95}]$ (min)**	**RR**	**Resistant status**
Ipono	4%DDT	Nov/2003	80	28.1 [22.4-31.9]	50.8 [45.0-62.7]	1.4	RC
	0.05%Deltamethrin		79	8.3 [5.4-10.9]	24.0 [18.1-38.9]	0.8	S
	0.75%Permethrin		79	11.5 [10.2-12.8]	44.3 [39.2-51.2]	1.3	S
	4%DDT	Nov/2007	87	>60	>60	>3	R
	0.05%Deltamethrin		95	13.7 [12.7-14.7]	34.1 [31.3-37.4]	1.4	RC
	0.75%Permethrin		88	37.3 [34.1-40.6]	>60	4.1	R
Nkolondom	4%DDT	Nov/2003	80	>60	>60	>3	R
	0.05%Deltamethrin		89	12.2 [3.7-18.8]	42.8 [32.7-61.0]	1.2	S
	0.75%Permethrin		86	12.6 [9.6-15.4]	47.1 [37.3-67.4]	1.4	S
	4%DDT	Nov/2007	84	>60	>60	>3	R
	0.05%Deltamethrin		89	20.1 [17.1-22.8]	45.8 [39. 1-58.4]	2.1	RC
	0.75%Permethrin		83	40.2 [37.6-42.8]	>60	4.5	RC
Mangoum	4%DDT	May/2003	80	>60	>60	>3	R
	0.05%Deltamethrin		84	24.1 [22.4-25.1]	39.2 [37.1-42.0]	2.5	R
	0.75%Permethrin		80	57.4 [52.3-68.0]	>60	6.6	R
	4%DDT	May/2007	91	>60	>60	>3	R
	0.05%Deltamethrin		84	19.4 [16.4-22.0]	48.3 [41.8-59.1]	2.0	RC
	0.75%Permethrin		106	54.2 [51.9-57.1]	>60	6.1	R
Bonanloka	4%DDT	May/2001	80	49.3 [41.8-58.7]	>60	2.6	RC
	0.05%Deltamethrin		86	22.4 [21.2-23.7]	34.8 [31.2-39.0]	2.4	RC
	0.75%Permethrin		91	16.3 [14.9-17.7]	25.7 [22.4-29.6]	1.7	R
	4%DDT	May/2005	80	51.6 [41.9-54.0]	>60	2.7	R

	0.05%Deltamethrin		83	8.3 [7.6-9.1]	18.6 [16.8-21.1]	0.8	S
	0.75%Permethrin		85	6.2 [3.6-8.5]	23.8 [18.0-37.2]	0.7	S
	4%DDT	May/2007	87	37.9 [36.5-39.5]	>60	1.9	RC
	0.05%Deltamethrin		84	12.8 [10.6-14.6]	24.1 [20.4-31.6]	1.3	RC
	0.75%Permethrin		88	15.2 [13.2-17.1]	35.3 [30.7-42.6]	1.7	RC

N: sample size; kdT50: knockdown time for 50% mosquitoes; kdT95: knockdown time for 95% mosquitoes; CI95: confidence interval at 95%; min: minutes; RR: resistance ratio (kdT50 of the tested population/kdT50of the Kisumu strain); S: susceptible; RC: resistance to be confirmed; R: resistant; Nov: November.

Nwane et al. BMC Infectious Diseases 2009 9:163, doi:10.1186/1471-2334-9-163

In the Ipono and Nkolondom populations, an increase of resistance to DDT and pyrethroids was observed from 2003 to 2007, as shown by an increase of knockdown times and a significant decrease in mortality ($p < 0{,}05$), particularly in the Ipono population (Table 2, Figure 2). Conversely, a decrease of resistance to DDT and pyrethroids was noted in the Mangoum population from 2003 to 2007; this decrease was mostly associated with significant increase of mortality rates ($p < 0.05$) in 2007 (Figure 2), although the knockdown times were slightly elevated. In Bonanloka, variations in resistance levels to DDT and pyrethroids were noted across the three surveys. DDT resistance was highest in 2005, whereas full susceptibility to pyrethroids was observed at this time. Reduced susceptibility to pyrethroids was observed in 2001 and 2007, however.

Mosquito Species, Molecular Forms and *Kdr* Frequencies

Samples of 40-50 specimens randomly drawn from control groups used for the susceptibility tests in each study site and period were identified molecularly to species and molecular form, and their genotype at the *kdr* locus was determined (Table 3). All mosquitoes tested belonged to the *An. gambiae* s.s species. The M and S molecular forms were found together in Ipono, Nkolondom and Bonanloka samples. The relative frequency of the M form which was 100% at Bonanloka in 2001 and 60.5% at Nkolondom in 2003 dropped to 58% and 4% in 2007, respectively ($p < 0.0001$). On the other hand, no significant difference was seen in the relative frequencies of both forms in Ipono between collections conducted in 2003 and 2007 ($\chi^2 = 0.17$, $p = 0.68$). In Mangoum, only the S molecular form was found throughout the 2 study periods.

Table 3: Variations in molecular forms and *kdr*-alleles frequencies in tested *Anopheles gambiae* populations

An. gambiae	Site	Year	N (%)	f(L1014S)	f(L1014F)
M-form	Ipono	2003	16 (33)	ND	0
		2005	22 (61)	0	0
		2007	19 (30)	0	0.10

	Nkolondom	2003	26 (60)	ND	0
		2005	0 (0)	-	-
		2007	2 (4)	0	0.50*
	Mangoum	2003	0 (0)	-	-
		2005	0 (0)	-	-
		2007	0 (0)	-	-
	Bonanloka	2001	30 (100)	ND	0
		2005	33 (63)	0	0
		2007	33 (58)	0	0.18
S-form	Ipono	2003	32 (67)	ND	0.04
		2005	14 (39)	0.03	0.25
		2007	45 (70)	0.18	0.62
	Nkolondom	2003	17 (39)	ND	0
		2005	64 (100)	0.03	0.59
		2007	45 (96)	0.02	0.92
	Mangoum	2003	51 (100)	ND	0.37
		2005	76 (100)	0.14	0.85
		2007	33 (100)	0.24	0.72
	Bonanloka	2001	0 (0)	-	ND
		2005	19 (37)	0	0.20
		2007	24 (42)	0	0.16

N: sample size; f(): frequency of the kdr alleles; *: 1 specimen out of the 2 tested carried the mutation; ND: not determined.

Nwane et al. BMC Infectious Diseases 2009 9:163, doi:10.1186/1471-2334-9-163

Both L1014F and L1014S *kdr* mutations were observed in three of the four *An. gambiae*populations tested, except in Bonanloka where only the L1014F mutation was recorded. Both mutations were found in individuals of the S molecular form while only the L1014F mutation was found in the M form, although at low frequencies (f < 0.2, Table 3) and in samples collected in 2007 only. In the three localities where the L1014S was detected in the S form, its frequency did not differ significantly between 2005 and 2007 (p > 0.07). However, a significant increase in the frequency of the L1014F mutation was detected in most of the S form populations between 2003 and 2007, especially in Nkolondom, where it rose from 0% in 2003 to 92% in 2007 (p < 0.0001).

Resistance Phenotype and *Kdr* Genotypes in M and S Molecular Forms from Ipono

To explore the distribution of resistance phenotypes and *kdr* genotypes within and between molecular forms of *An. gambiae*, we further analysed the samples from Ipono assayed in 2007, because of presence of both M and S molecular forms, and of both L1014F and L1014S *kdr*mutations in this site. A total of 214 *An. gambiae* s.s. specimens, randomly selected from control, dead or survivors to susceptibility tests (20-65 specimens per test) were successfully analysed. Among these mosquitoes, 46 (21.5%) were M form and 168 (78.5%) were S form (Table 4). No M form mosquito was identified within the group of survivors to insecticide exposure, suggesting complete susceptibility to the three compounds. Some dead specimens carried the L1014F allele at the heterozygous or homozygous state.

Table 4: Phenotypes and genotypes at codon 1014 of the vgsc of *Anopheles. gambiae*collected in Ipono (December 2007)

Molecular form	Phenotype	Insecticide	N	Genotypes						f(Ser)	f(Phe)
				Leu-Leu	Leu-Phe	Leu-Ser	Ser-Ser	Phe-Phe	Phe-Ser		
M-form	Control	-	19	15	4	0	0	0	0	0	0.10
	Dead	4%DDT	8	6	2	0	0	0	0	0	0.12
		0.05%Deltamethrin	6	6	0	0	0	0	0	0	0
		0.75%Permethrin	13	8	2	0	0	3	0	0	0.31
	Survivor	4%DDT	0	-	-	-	-	-	-	-	-
		0.05%Deltamethrin	0	-	-	-	-	-	-	-	-
		0.75%Permethrin	0	-	-	-	-	-	-	-	-
S-form	Control	-	45	4	7	2	3	20	9	0.18	0.62
	Dead	4%DDT	11	2	4	3	0	1	1	0.18	0.41
		0.05%Deltamethrin	29	2	3	3	2	18	1	0.14	0.69
		0.75%Permethrin	15	2	4	1	1	4	3	0.20	0.50
	Survivor	4%DDT	29	0	1	0	0	27	1	0.01	0.96
		0.05%Deltamethrin	6	0	0	0	0	3	3	0.25	0.75
		0.75%Permethrin	33	0	0	0	0	24	9	0.14	0.86′

N: sample size; f (): frequency of the resistant alleles (L1014S = Ser and L1014F = Phe) in the mosquito pools; vgsc: voltage gated sodium channel

Nwane et al. BMC Infectious Diseases 2009 9:163, doi:10.1186/1471-2334-9-163

Among the 168 individuals of the S form, 26.8% belonged to control, 32.7% to dead and 40.5% to survivor samples. The two *kdr* mutations were present in control, dead or survivor mosquitoes, with a considerable number of homozygote (L1014F/L1014F) and heterozygote (L1014F/L1014S) individuals in each of the three mosquito classes. All possible genotype combinations were observed in the control and dead mosquitoes, whereas only three genotypes were recorded in surviving mosquitoes, most of which were either double heterozygotes (L1014F/L1014S) or homozygotes for the "West African" *kdr* mutation (L1014F/L1014F). No significant difference was seen in the frequency of the L1014S allele between dead and survivors to susceptibility tests, except with DDT where the frequency was lower in survivors ($p < 0.001$). The frequency of L1014F allele was significantly higher in mosquitoes surviving to DDT and permethrin (0.86-0.96) than in dead mosquitoes (0.41-0.50, $P < 0.001$). On the other hand, no significant difference was seen in L1014F frequency between dead and survivors to deltamethrin ($P = 0.34$). Overall, the frequencies of L1014F allele were higher in survivors than in the two other classes of mosquitoes.

DISCUSSION

The survey on insecticide usage provided evidence for intensive use of chemical insecticides both for agro-industry and for personal protection in South Cameroon. It is consistent with observations made by Akogbéto and colleagues [46] that the choice and usage of chemical compounds depends on cultivated crops and specification of devastating insects.

Susceptibility data presented above confirmed that wild *An. gambiae* s.s. from the study area show reduced susceptibility to DDT and pyrethroids [34,36, 47]. The level of resistance however, varies from one year to another, according to the proportion of the resistant S form in the populations where the two molecular forms of *An. gambiae* were found together, especially in Nkolondom. In addition, the frequency of both *kdr* alleles increased in time within the S form in all sites where it was found. These results testify that the spread of the *kdr* alleles is an ongoing process in *An. gambiae* mosquito populations from Cameroon [36], as well as elsewhere in Central Africa [48,49].

The presence of the *kdr* mutations in the malaria vector *An. gambiae* has been monitored using a variety of molecular techniques [50]. In this study, molecular detection of East and West *kdr*genotypes were conducted using two PCR methods differing in their performance. Although the protocol assay used in 2003 [28] could not detected the 1014S allele, data collected in 2005 and 2007 using the protocol described by Lynd and colleagues [44] demonstrate the rapid invasion of the *kdr* mutation in the studied *An. gambiae* populations, especially in Ipono and Nkolondom.

Evidence of a strong correlation between resistance phenotype and *kdr* genotypes has been documented in previous studies [28,29]. However, results provided in this study, agree with the standpoint suggested by Brooke [50], arguing that *kdr* may act with certain co-factors that are thus far unidentified. This resistance mechanism could be multigenic, and the *kdr* genotype might not fully explain all the variance in the resistance phenotype [51]. However, this hypothesis does not rule out the causal relationship between the *kdr* genotype and susceptibility to DDT and pyrethroids [52].

The higher number of mutant homozygotes L1014F recorded in dead mosquitoes when exposed to 0.05% deltamethrin demonstrates that, *kdr* may be less efficient in providing resistance to deltamethrin than it is for DDT and permethrin resistance, as previously suggested [37,53,54]. Alternatively, the diagnostic dosage of deltamethrin may be very high and therefore killing resistant mosquitoes. Furthermore, other resistance mechanisms such as elevated oxidase, esterase or glutathione S-transferase activity may add to the *kdr* effect and increase resistance to DDT and permethrin in Ipono. Metabolic resistance has already been reported in *An. gambiae* and*An. arabiensis* populations from Cameroon [25,35,55], although specific analysis were not carried out to explore these mechanisms within each molecular form. Meanwhile, resistance levels such as that reported here seem to be strongly associated with the S molecular form. However, attention must also be paid to the M molecular form considering the increasing frequency of the L1014F *kdr*allele in this form [49].

The rise in frequency of both *kdr* mutations in *An. gambiae* is probably facilitated by the intensive use of the same insecticides in agriculture, wood industry and public health, resulting in ubiquitously high selection pressure for resistance in insects [56]. Reports from Cameroon as well as other African countries have mentioned the relation between public health or agricultural use of insecticides and the evolution of insecticide resistance in malaria vectors [34,55-57]. In addition to cotton or rice cultivation areas, market gardening areas, public health or personal protection, the current study put forward forest exploitation sites (timber yards) as potential zones of insecticide resistance emergence in *An. gambiae*.

This study further highlights the dynamics of *An. gambiae* susceptibility to insecticides in southern Cameroon. Biological, genetic and environmental factors may intervene in modulating susceptibility to insecticides.

CONCLUSIONS

This study underlines the variability in the resistance level of *An. gambiae* s.s from southern Cameroon to the first-line insecticides currently used for its control. The current distribution of *kdr*mutations and ongoing trends for their geographical spread and rise in frequency need to be carefully considered. Such a situation is a real threat for malaria vector control strategies implemented in South Cameroon and calls for close collaboration between pest-management and vector control programmes towards the implementation of sustainable resistance management strategies.

AUTHORS' CONTRIBUTIONS

JE and FS conceive the study. JE, FS, RM, and PN designed the study protocol; JE, PN, MC, JCT, CKH and HPAA performed field work and bioassays; PN performed molecular analyses; PN and JE analyzed and interpreted the data; PN drafted the manuscript which was critically revised by JE, RM and FS. All the authors read and approved the final manuscript.

ACKNOWLEDGEMENTS

We wish to thank Léonard Enama Nga, Nicolas Enoga and the communities in our study sites for their help and collaboration in the fields.

Financial support was provided by UNICEF/UNDP/World Bank/WHO/Special Programme for Research and Training in Tropical Diseases (TDR): Multilateral Initiative on Malaria grant N° A10625 and Re-entry Grant number A41481 to J.E.

REFERENCES

1. Gariou J, Mouchet J: Apparition d'une souche d'*Anopheles gambiae* résistante à la dieldrine dans la zone de campagne antipaludique du Sud-Cameroun. *Bull Soc Path Exot* 1961, 54:870-875.
2. Livadas G, Mouchet J, Gariou J, Chastang R: Peut-on envisager l'éradication du paludisme dans la region forestière du Sud-Cameroun? *Revista di Malariologia* 1958, 37:229-256.
3. Trigg PI, Kondrachine V: *Bull World Health Organisation*. 1998, 76:11-16.
4. Cavalié P, Mouchet J: Les campagnes expérimentales d'éradication du paludisme dans le Nord de la République du Cameroun. *Med Trop* 1962, 22:95-118.
5. Mittal PK, Adak T, Sharma VP: Acute toxicity of certain organochlorine, organophosphorus, synthetic pyrethroid and microbial insecticides to the mosquito fish *Gambusia affinis* (Baird and Girard). *Indian J Malariol* 1991, 28:167-70.
6. Zaim M, Aitio A, Nakashima N: Safety of pyrethroid-treated nets. *Med Vet Entomol* 2000, 14:1-5
7. WHO: Global strategic framework for integrated vector management. *WHO/CDS/CPE/PVC/2004.10. Geneva, Switzerland*
8. WHO: Malaria vector control and personal protection. *WHO Technical Report Series N° 936* 2006

9. Guillet P: La lutte antivectorielle. *Bull Soc Pathol Exot* 2001, 94(2 bis):165-168

10. Coleman M, Hemingway J: Insecticide resistance monitoring and evaluation in disease transmitting mosquitoes. *J Pest Sci* 2007, 32(2):69-76

11. Georghiou GP: *"Managing Resistance to Agrochemicals: From Fundamental Research to Practical Strategies.* Edited by Green MB, Lebaron HM, Moberg WK. American Chemical Society; Washington; 1990:18-41.

12. Green MB, LeBaron HM, Moberg WK: *Managing resistance to agrochemicals.* ACS Symposium Series; Washington DC; 1990:3-16.

13. Hargreaves K, Koekemoer LL, Brooke BD, Hunt RH, Mthembu J, Coetzee M: *Anopheles funestus* resistant to pyrethroid insecticides in South Africa. *Med Vet Entomol* 2000, 14:1-9

14. Kazachkova NI: Genotype analysis and studies of pyrethroid resistance of the oilseed rape (*Brassica napus*) insect pest-pollen beetle (*Meligethes aeneus*). In *Doctoral Thesis*. Swedish University of Agricultural Sciences, Sweden; 2007:56.

15. Ranjekar PK, Aparna P, Gupta VB, Raj JB, Ananda KP: Genetic engineering of crop plants for insect resistance. *Current Science* 2003, 84(3):321-329.

16. N'Guessan R, Corbel V, Akogbéto M, Rowland M: Reduced efficacy of insecticide-treated nets and indoor residual sprying for malaria control in pyrethroid resistance area, Benin. *Emerg Infect Dis* 2007, 13:199-206.

17. Sachs J, Malaney P: The economic and social burden of malaria. *Nature* 2002, 415:680-685.

18. Oppenoorth FJ: Biochemistry and genetics of insecticide resistance. In *Comprehensive insect physiology. Pergamon, Oxford* Edited by Kerkut GA, Guilbert LL. 1985, 731-773.

19. Prapanthadara L, Hemingway J, Ketterman AJ: Partial purification and characterization of glutathione S-transferases involved in DDT resistance from the mosquito*Anopheles gambiae*. *Pest Biochem Physiol* 1993, 47:119-133

20. Chandre F, Darriet F, Manguin S, Brengues C, Carnevale P, Guillet P: Pyrethroid cross-resistance spectrum among populations of

Anopheles gambiae s.s from Côte d'Ivoire. *J Am Mosq Control Ass* 1999, 15:53-9.

21. Hemingway J, Hawkes NJ, Mc Carroll L, Ranson H: The molecular basis of insecticides resistance in mosquitoes. *Insect Biochem Molec Biol* 2004, 34:653-65
22. Poirié M, Raymond M, Pasteur N: Identification of two distinct amplifications of the esterases B locus in *Culex pipiens* (L) Mosquitoes from Mediterranean countries. *Biochem Genet* 1992, 30:13-26
23. Raymond M, Chevillon C, Guillemaud T, Leonormand T, Pasteur N: An overview of evolution of the overproduced esterases in the mosquito *Culex pipiens*. *Philos Trans R Soc London B Biol Sci* 1998, 353:1707-1711
24. Rooker S, Guillemaud T, Bergé J, Pasteur N, Raymond M: Co-amplification of esterases A and B genes as a single unit in *Culex pipiens* mosquitoes. *Heredity* 1996, 77:555-561
25. Müller P, Chouaibou M, Pignatelli P, Etang J, Walker ED, Donnelly MJ, Simard F, Ranson H:Pyrethroid tolerance associated with elevated expression of antioxidants and agricultural practice in *Anopheles arabiensis* sampled from an area of cotton fields in Northern Cameroon. *Mol Ecol* 2008, 17(4):1145-1155
26. Müller P, Donnelly MJ, Ranson H: Transcription profiling of a recently colonized pyrethroid resistant *Anopheles gambiae* strain from Ghana. *BMC Genomics* 2007, 8:36.
27. Brogdon G, Mc Allister J: Insecticide resistance and vector control. *Emerg Infect Dis* 1998, 4:605-613.
28. Martinez-Torres D, Chandre F, Williamson MS, Darriet F, Bergé JB, Devonshire AL, Guillet P, Pasteur N, Pauron D: Molecular characterization of pyrethroid knockdown resistance (*kdr*) in the major malaria vector *Anopheles gambiae* s.s. *Insect Mol Biol* 1998, 7:179-184
29. Ranson H, Jensen B, Vulvule JM, Wang X, Hemingway J, Collins FH: Identification of a point mutation in the voltage-gated sodium channel gene of Kenyan *Anopheles gambiae* associated with resistance to DDT and pyrethroids. *Insect Mol Biol* 2000, 9:491-497

30. Walsh SB, Dolden TA, Moores GD, Kristensen M, Lewis T, DevonshireAL, Williamson SM:Identification and characterization of mutations in housefly (*Musca domestica*) acetylcholinesterase involved in insecticide resistance. *Biochem J* 2001, 359:175-181

31. Weill M, Fort P, Berthomieu A, Dubois MP, Pasteur N, Raymond M: A novel acetylcholinesterase gene in mosquitoes codes for the insecticide target and is non-homologous to the ace gene in *Drosophila*. *Proc R Soc* 2002, 269:2007-2016

32. N'Guessan R, Darriet F, Guillet P, Carnevale P, Traoré-Lamizana P, Corbel V, Koffi AA, Chandre F: Resistance to carbosulfan in field populations of *Anopheles gambiae*from **Côte d'Ivoire based on reduced sensitivity of acetylcholinesterase.** *Med Vet Entomol* 2003, 17:19-25

33. Djogbénou L, Dabiré R, Diabaté A, Kengne P, Akogbéto M, Hougard JM, Chandre F:Identification and geographic distribution of the Ace-1^R mutation in the malaria vector *Anopheles gambiae* in South-Western Burkina Faso, West Africa. *Am J Trop Med Hyg* 2008, 78(2):298-302

34. Etang J, Manga L, Chandre F, Guillet P, Fondjo E, Mimpfoundi R, Toto JC, Fontenille D:Insecticide susceptibility status of *Anopheles gambiae s.l.* (Diptera: Culicidae) in the Republic of Cameroon. *J Med Entomol* 2003, 40:491-497.

35. Etang J, Manga L, Toto JC, Guillet P, Fondjo E, Chandre F: Spectrum of metabolic-based resistance to DDT and pyrethroids in *Anopheles gambiae* s.l populations from Cameroon. *J Vect Ecol* 2007, 32(1):123-133

36. Etang J, Fondjo E, Chandre F, Brengues C, Nwane P, Chouaúbou M, Ndjemai H, Simard F:First report of knockdown mutations in the malaria vector *Anopheles gambiae* from Cameroon. *Am J Trop Med Hyg* 2006, 74:795-797.

37. Reimer L, Fondjo E, Patchoké S, Diallo B, Lee Y, Arash NG, Ndjemai HM, Atangana J, Traoré SF, Lanzaro G, Cornel JA: Relationship between *kdr* mutation and resistance to pyrethroid and DDT insecticides in natural populations of *Anopheles gambiae*. *J Med Entomol* 2008, 45(2):260-266.

38. Suchel JB: *Les climats du Cameroun.* Thèse de Doctorat d'Etat. Université de Bordeau III; 1987:1186.

39. Gillies MT, De Meillon B: *The anophelinae of Africa South of the Sahara.* second edition. South African Institute of Medical Research, Johannesburg; 1968.

40. Gillies MT, Coetzee M: *Supplement. The anophelinae of Africa South of the Sahara.* South African Institute of Medical Research, Johannesburg; 1987.

41. WHO: Test procedures for insecticide resistance monitoring in malaria Vectors. Bio-efficacy and Persistence of insecticides on treated surfaces. *WHO/MAL/98.,12. Report of the WHO Informal Consultation, Geneva* 1998, 43.

42. Morlais I, Ponçon N, Simard F, Cohuet A, Fontenille D: Intraspecific nucleotide variant in*Anopheles gambiae*: new insights into biology of malaria vectors. *Am J Trop Med Hyg* 2004, 71(6):795-802

43. Fanello C, Santolamazza F, della Torré A: Simultaneous identification of species and molecular forms of the *Anopheles gambiae* complex by PCR-RFLP. *Med Vet Entomol* 2002, 16:461-464

44. Lynd A, Ranson H, McCall PJ, Randle NP, Black WC IV, Walker ED, Donnelly MJ: A simplified high-throughput method for pyrethroid knockdown resistance (*kdr*) detection in *Anopheles gambiae*. *Malaria J* 2005, 4:16

45. Finney DJ: *Probit analysis.* 3rd edition. Cambridge University Press; Cambridge; 1971.

46. Akogbéto MC, Djouaka R, Noukpo H: Utilisation des insecticides agricoles au Bénin. *Bull Soc Pathol Exot* 2005, 98(5):400-405

47. Ndjemaï HNM, Patchoké S, Atangana J, Etang J, Simard F, Bilong Bilong CF, Reimer L, Cornel A, Lanzaro CG, Fondjo E: The distribution of insecticide resistance in *Anopheles gambiae* s.l populations from Cameroon: an update. *Trans Roy Soc Trop Med Hyg* 2008, in press.

48. Pinto J, Lynd A, Elissa N, Donnelly MJ, Costa C, Gentile G, Caccone A, Do Rosário VE: Co-occurrence of East and West African *kdr* mutations suggests high levels of resistance to

pyrethroid insecticides in *Anopheles gambiae* from Libreville, Gabon. *Med Vet Entomol* 2006, 20:27-32

49. Santolamazza F, Calzetta M, Etang J, Barrese E, Dia I, Caccone A, Donnelly MJ, Petrarca V, Simard F, Pinto J, della Torre A: Distribution of knockdown resistance mutations in*Anopheles gambiae* molecular forms in West and West-Central Africa. *Malar J* 2008, 7:74
50. Bass C, Nikou D, Donnelly MJ, Williamson MS, Ranson H, Ball A, Vontas J, Field LM:Detection of knockdown resistance (*kdr*) mutations in *Anopheles gambiae*: a comparison of two new high-throughput assays with existing methods. *Malaria J* 2007, 6:111.
51. Brooke BD: *Kdr*: can a single mutation produce an entire insecticide resistance phenotype? *Trans Roy Soc Trop Med Hyg* 2008, 102:524-525.
52. Donnelly MJ, Corbel V, Weetman D, Wilding CS, Williamson MS, Black WC IV: Does *kdr*genotype predict insecticide-resistance phenotype in mosquitoes. *Trends in Parasitology* 2009, 25(5):213-9.
53. Awolola TS, Oduola AO, Oyewole IO, Obansa JB, Amajoh CN, Koekemoer LL, Coetzee M:Dynamics of knockdown pyrethroid insecticide resistance alleles in a field population of *Anopheles gambiae* s.s in southwestern Nigeria. *J Vect Borne* 2007, 44:181-188.
54. Abdalla H, Matambo TS, Koekemoer LL, Mnzava AP, Hunt RH, Coetzee M: Insecticide susceptibility and vector status of natural populations of *Anopheles arabiensis* from Sudan. *Trans Roy Soc Trop Med Hyg* 2008, 102:263-271
55. Himeidan YE, Chen H, Chandre F, Donnelly MJ, Yan G: Permethrin and DDT resistance in the malaria vector *Anopheles arabiensis* from Eastern Sudan. *Am J Trop Med Hyg* 2007, 77:1066-1068
56. Diabaté A, Baldet T, Chandre F, Akogbéto M, Guiguemde RT, Darriet F, Brengues C, Guillet P, Hemingway J, Graham JS, Hougard JM: The role of agricultural use of insecticides in resistance to pyrethroids in *Anopheles gambiae s.l.* in Burkina Faso. *Am J Trop Med Hyg* 2002, 67:617-622.
57. Chouaïbou M, Etang J, Brevault T, Nwane P, Hinzoumbé CK, Mimpfoundi R, Simard F:Dynamics of insecticide resistance

in the malaria vector *Anopheles gambiae* s.l from an area of extensive cotton cultivation in Northern Cameroon. *Trop Med Int Hlth* 2008, 13(4):1-11.

Chapter 2

Acetylcholinesterase Inhibition by Biofumigant (coumaran) from Leaves of Lantana Camara in Stored Grain and Household Insect Pests

Yallappa Rajashekar[1], Anjanappa Raghavendra[2], and Nandagopal Bakthavatsalam[2]

[1]Animal Bioresources Programme, Institute of Bioresources and Sustainable Development, Department of Biotechnology, Government of India, Takyelpat, Imphal, Manipur 795001, India

[2]Division of Insect Ecology, National Bureau of Agriculturally Important Insects, Bangalore, Karnataka 560024, India

ABSTRACT

Recent studies proved that the biofumigants could be an alternative to chemical fumigants against stored grain insect pests. For this reason, it is necessary to understand the mode of action of biofumigants. In the present study the prospectus of utilising Lantana camara as a potent fumigantinsecticideisbeingdiscussed. Inhibitionofacetylcholinesterase (AChE) by Coumaran, an active ingredient extracted from the plant L. camara, was studied. The biofumigant was used as an enzyme inhibitor and acetylthiocholine iodide as a substrate along with Ellman's reagent to carry out the reactions. The in vivo inhibition was observed in both dose dependent and time dependent in case of housefly, and the nervous tissue (ganglion) and the whole insect homogenate of stored grain insect exposed to Coumaran. The possible mode of action of Coumaran as an acetylcholinesteraseinhibitor is discussed.

INTRODUCTION

Acetylcholine (Ach) is one of the major molecules by which nerve impulses are transmitted from nerve cell or involuntary muscle [1]. Acetylcholinesterase AChE is an enzyme that breaks down the neurotransmitter acetylcholine at the synaptic cleft (the space between two nerve cells) so the next nerve impulse can be transmitted across the synaptic gap [2]. The phosphine, organophosphates, and carbamates act by interfering with the passage of impulses in the insect nervous system [3]. Organophosphate insecticides are generally regarded as irreversible inhibitors of the enzyme acetylcholinesterase. The inability of phosphorylated AChE to hydrolyse acetylcholine, the buildup of concentration of the acetylcholine in the synapse and excessive neuro excitation are the results of prolonged binding of ACh to its postsynaptic receptor. The signs of intoxication include restlessness, hyperexcitability, tremors, convulsions and paralysis leading to death [4, 5].

Stored grain insects cause extensive damage in stored wheat, rice, pulses and other commodities [6, 7]. Synthetic insecticides especially fumigants such as methyl bromide and aluminium phosphide are commonly used to manage these pests. Even though this method is effective, repeated use of these chemicals may cause environmental hazards and various biochemical changes in nontarget animals [8]. It was also reported that the insects have developed resistance against these fumigants [9]. Due to the possible role of methyl bromide in depletion of ozone layer, its use as a fumigant is banned [10]. Thus, there is a need to develop cheaper and safer alternative measures including plant based products against stored grain and household insect pests [11–13].

In this perspective, properties of plant products including essential oils and their bioactive molecules have been broadly studied for the control of stored grain pests [1, 14–16]. The properties include toxic, ovicidal, repellent, antifeedant and other properties [17]. Essential oils, allelochemicals and their individual constituents have been known to play an important role as protectants of stored grains and proved to possess repellent and insecticidal properties [18]. The different organic extracts from root powder of Decalepis hamiltonii (Wight and Arn) and the bioactive compounds from Decalesides showed potential to be used as grain protectants against grain insect pests [14, 19]. Many plant secondary metabolites such as monoterpenoids, polyphenols, and sugars have insecticidal activity against stored grain insect pests [7, 20]. However, the detailed studies on the biochemical effects of these compounds on insect physiology would reveal the affected target sites in the respective pests.

Several reports indicate that monoterpenoids and most of the plant volatiles cause insect mortality by inhibiting acetylcholinesterase enzyme [1]. The monoterpenoid (1, 8-cineole) was reported to be potent AChE inhibitor [21]. The action of essential oils and biofumigants on insects could be neurotoxic based on behavioural symptoms similar to those produced by organophosphates [22] or inhibition of AChE [23]. Evidently, the studies on mode of action of biofumigants in insects have largely focused on monoterpenoids. So, it becomes very important to study the mode of other compounds showing fumigant action.

Lantana camara (Verbenaceae), an erect shrub, grows widely throughout the tropical, subtropical and temperate parts of the world. Earlier work showed that the leaves of L. camara were the source of insecticidal activity [24]. Previously, we investigated the insecticidal activity of extracts from leaves of L. camara against the storage pests, namely, rice weevil, Sitophilus oryzae L. (Coleoptera: Curculionidae) rust red flour beetle, Tribolium castaneum Herbst. (Coleoptera:

Tenebrionidae), and adzuki bean weevil, Callosobruchus chinensisFab. (Coleoptera: Bruchidae) [25]. the present study aims to investigate the mode of action of Coumaran in the inhibition of acetylcholinesterase.

MATERIALS AND METHODS

Insects

Cultures of S. oryzae were maintained on whole wheat (Triticium aestivum), whereas T. castaneum were reared on whole wheat flour with 2% yeast powder. Houseflies (Musca domestica) were reared in a mixture of sterilized bran, milk powder and water. The adults were allowed free access to water and thick paste of condensed milk and milk powder. Cultures were maintained at 30 ±1°C and 70% relative humidity. Adult's ofS. oryzae (3–5 d) and other species (2-3 d) were used for the experiments. Experiments were carried out in the laboratory at 27±2°C and 70±5% r.h. [14, 25].

Chemicals

AChE, acetylthiocholine iodide, Coumaran, and Pyridostigmine bromide were procured from Sigma chemical Co. (St. Louis, MO, USA). Ammonium molybdate, ascorbic acid, trichloroacetic acid (TCA), hydrochloric acid (HCl) and other chemicals were procured from Sisco Research Laboratory, Mumbai, India.

Biofumigant Preparation

Healthy, mature leaves of L. camara were collected from Hesaraghatta, Bangalore, India. The leaves were cut into small pieces, shade-dried and powdered. One hundred gram of leaf powder was sequentially extracted with a series of solvents of increasing polarity, namely, hexane, ethyl acetate, acetone and methanol in a Soxhlet apparatus. The solvent was evaporated in vacuo and the residue was dissolved in a known volume of methanol. This solution was screened for insecticidal activity using contact/fumigant toxicity bioassay. Since the methanolic extract was found to show the highest insecticidal activity in the preliminary screening, this was selected for the isolation of the bio fumigant [25].

Inhibition of AChE

In vivo inhibition of AChE in relation to the toxicity of Coumaran was investigated in both houseflies and stored grain pests.

Dose-Response Study. Insects were exposed to LC_{25}, LC_{50}, and LC_{90} doses of Coumaran in the fumigant bioassay [25]. The doses were selected based on the results of toxicity of Coumaran to stored grain insects and housefly. Respective batches of solvent treated controls were also employed. After 45 min exposure insects were removed. In case of houseflies the head and thorax were dissected out and frozen for enzyme assay, where as in case of stored grain insects, whole insect was homogenized and stored at −20°C for enzyme assay.

Time-Course Study. Insects were exposed to single LC_{50} dose of Coumaran in the fumigation bioassay and removed at various exposure time intervals (15, 30, 45, and 60 min). Solvent (ethanol) treated groups served as control. After various time intervals of exposure, insects were removed and the insect tissues were dissected as described above for enzyme assay.

Inhibition of acetylcholinesterase (AChE) was estimated by colorimetric method previously described by Ellman et al. [26]. The reaction was carried out in a cuvette by dissolving Coumaran in absolute ethanol to which 40 μL of substrate (acetylthiocholine iodide) was added and then DTNB (200 μL) followed by inhibitor (Pyridostigmine bromide) solution (1 mL) in different concentrations of 0.01 μM, 0.1 μM, 0.5 μM, and 1 μM. Tests and control assays (without Coumaran) were corrected by blanks for nonenzymatic hydrolysis. Each assay was replicated only twice since the results from the second replication were practically similar to those produced by the first replication. Level of AChE activity was estimated by using Shimadzu UV-1600 Spectrophotometer at 412 nm and at 25°C.

In Vitro Inhibition of Acetylcholinesterase

In in vitro inhibition of acetylcholinesterase by Coumaran in the head and thorax of houseflies, whole insect homogenate of stored grain insects was studied. The enzyme was preincubated with Coumaran (0.01 μM–1 μM) at 37°C for 30 min and the inhibition of AChE was determined. LC_{50} were calculated by regression analysis.

Protein content was measured by the method of Lowry et al. [27] using BSA as the standard.

Statistical Analysis

The data obtained from the studies was analysed using one-way ANOVA at ($P<0.01$) and mean values were separated by using Tukey and Statplus 2007 software. The data was expressed as mean ±SD. Probit analysis was performed for calculatingLC_{25}, LC_{50} and, LC_{90} [28].

RESULTS

In Vivo Inhibition of Acetylcholinesterase in Relation to Insect Toxicity

Dose Response. Activity of acetylcholinesterase in insects exposed to LC_{25}, LC_{50}, and LC_{90} doses of Coumaran was markedly inhibited in dose-dependent studies in the head and thorax of housefly (Figure 1) and homogenate of stored grain insect pests (Figure 2). The in vivo enzyme inhibition was dose dependent and correlated with the knock-down effect measured at 45 min of exposure in the fumigation bioassay.

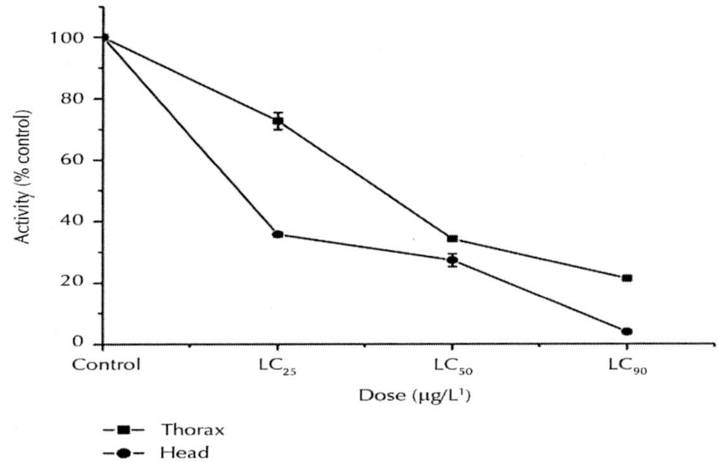

Figure 1: Dose-dependent inhibitions in vivo of AchE in the housefly by Coumaran at 45 min of exposure by fumigation (control activity: head homogenate = 178.03 micromoles of substrate hydrolyzed/minute/mg of protein; thorax homogenate = 169.14 micromoles of substrate hydrolyzed/minute/mg of protein) (n=6, error bars, standard deviation).

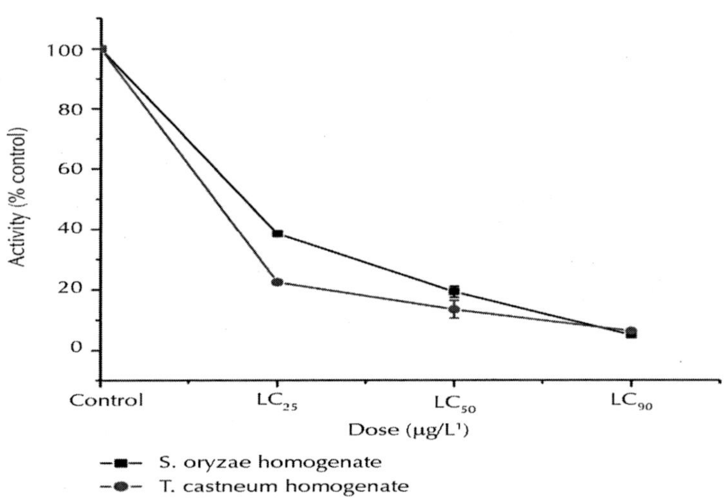

Figure 2: Dose-dependent inhibitions in vivo of AchE in the homogenate of S. oryzae andT. Castaneum by Coumaran at 45 min of exposure by fumiga-

tion (control activity: S. oryzae homogenate = 314.6 micromoles of substrate hydrolyzed/minute/mg of protein; T. castaneum homogenate = 348.8 micromoles of substrate hydrolyzed/minute/mg of protein) (n=6, error bars, standard deviation).

Time-Course Study. In houseflies treated with a single LC_{50} dose of Coumaran at various exposure times (0–45 min), inhibition of acetylcholinesterase increased with time and correlated with the knockdown effect (Figures3 and 4).

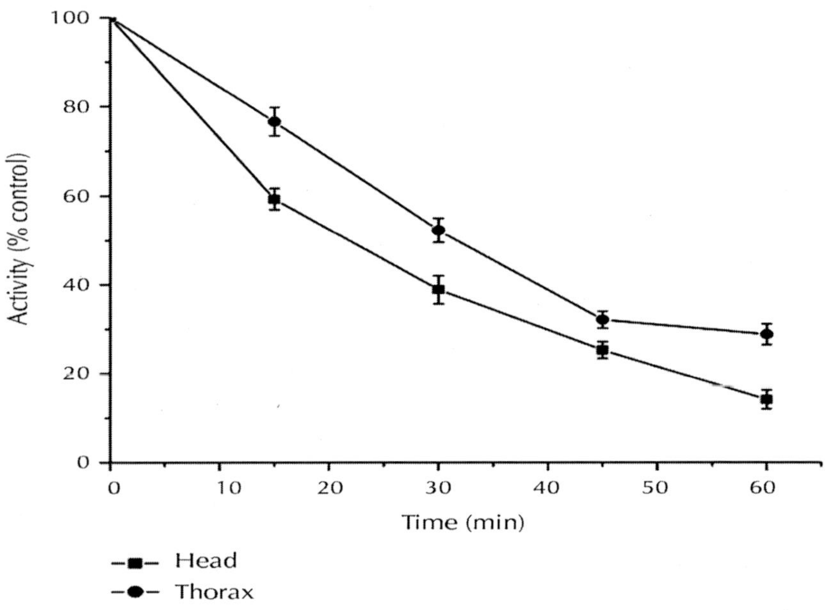

Figure 3: Time-course of in vivo inhibition of AchE in relation to knockdown of houseflies treated with Coumaran at LC_{50} concentration (0.032 μg/L) by fumigation (control activity: head homogenate = 188.7 micromoles of substrate hydrolyzed/minute/mg of protein; thorax homogenate = 166.5 micromoles of substrate hydrolyzed/minute/mg of protein) (n=6, error bars, standard deviation).

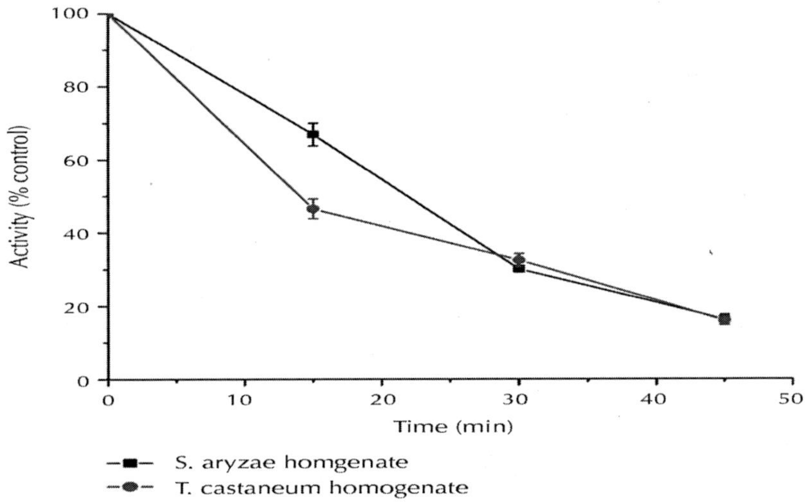

Figure 4: Time-course of in vivo inhibition of AchE in relation to knock-down of stored grain beetles treated with Coumaran at LC_{50} concentration (0.032 μg/L) by fumigation (control activity: S. oryzae homogenate = 304.6 micromoles of substrate hydrolyzed/minute/mg of protein; T. castaneum homogenate = 328.2 micromoles of substrate hydrolyzed/minute/mg of protein) (n=6, error bars, standard deviation).

In Vitro Inhibition of Acetylcholinesterase

The in vitro inhibition of acetylcholinesterase was increased with the concentration of Coumaran in the housefly (head and thorax) homogenate (Figures 5 and 6).

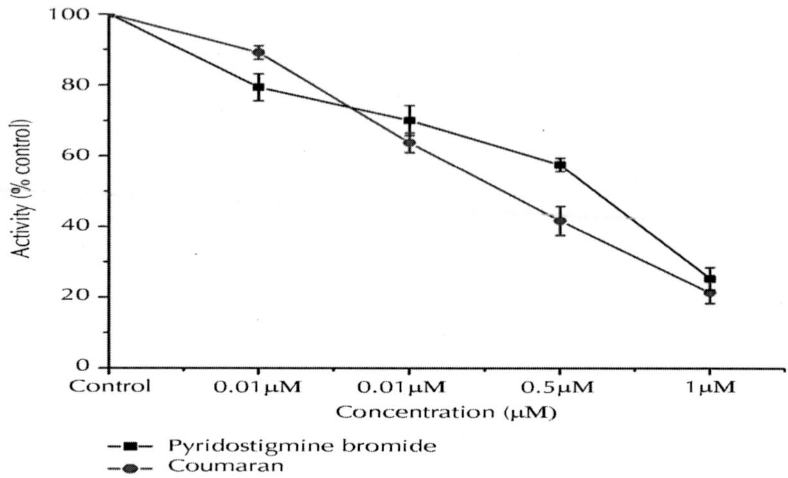

Figure 5: In vitro inhibition of AchE by Coumaran in comparison with that of Pyridostigmine bromide in houseflies (control activity: head homogenate = 238.3 micromoles of substrate hydrolyzed/minute/mg of protein) (n=6, error bars, standard deviation).

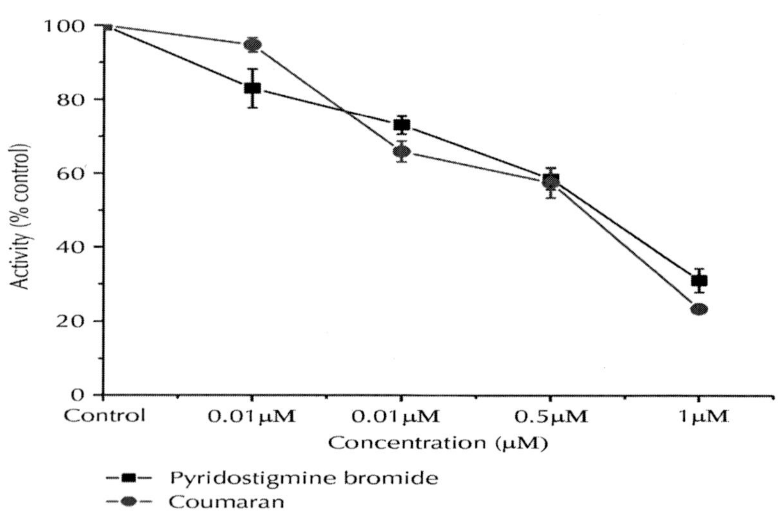

Figure 6: In vitro inhibition of AchE by Coumaran in comparison with that of Pyridostigmine bromide in houseflies (control activity: thorax homogenate =

249.1 micromoles of substrate hydrolyzed/minute/mg of protein) (n=6, error bars, standard deviation).

DISCUSSION

Due to environmental concerns, health hazards to human beings and the development of resistance in insect pests to the various insecticides, constant efforts were made to discover newer insecticides both from natural sources and by chemical synthesis. The majority of insecticides have mode of action that can be classified into three categories, namely, neurotoxins, respiratory inhibitors and growth regulators [29, 30]. Most of the chemical insecticides involve target sites in the nervous system, namely, acetylcholinesterase (organophosphate and carbamate compounds), voltage-gated sodium channels (pyrethroids and organochlorines), and the acetylcholine receptor (neonicotinoids) [31–33]. Although, insect control agents (e.g., azadirachtin, JH analogues and ecdysone antagonists) acting on hormone system such as insect growth regulators have been developed, due to their lack of contact and fumigant toxicity they are not highly successful, however they do found a place in integrated pest management scheme.

Although our experimental evidence indicates that the action of Coumaran involves the respiratory pigments possibly acting on the spiracles, the insects became hyperactive indicating neural excitation which is followed by knockdown effect and mortality which remains puzzling. The symptoms excitation, respiratory spiracles and knockdown of the insects upon fumigation, point out the affected targets as neural/neuromuscular sites. However, we investigated the possible involvement of two biochemical targets critical in the transmission of nerve impulse, namely, the enzymes, acetylcholinesterase and Na^+, K^+ ATPase. Acetylcholinesterase (AChE), involved in the synaptic transmission of nerve impulse, is the target for insecticides belonging to the organophosphorus and carbamate group [31, 33]. Preliminary experiments showed no effect of Coumaran on Na^+, K^+ ATPase in insects exposed to Coumaran (data not shown).

Present study showed that AChE is inhibited severely in insects (both housefly and stored grain insects) exposed to Coumaran in the fumigation bioassay. This in vivo inhibition closely correlates ($r \geq 0.9$) with the insect toxicity of Coumaran in dose-response and time-course studies. The in vivo inhibition was seen in both head and thorax in the case of housefly and the nervous tissue (ganglion) and whole insect homogenate of stored grain insect exposed to Coumaran.

Our results showed that the insects exposed to Coumaran are initially hyperactive indicating neural excitation, which is followed by knockdown effect, a symptom similar to that of insects exposed to carbamate suggesting a neurotoxic effect. The fundamental difference, however, is that the Coumaran action is mediated by fumigation with respiratory spiracles, unlike that of carbamate which acts by contact at any point of the body surface of insects. It is known that carbamates act by interfering inhibition of AChE; the carbamylated enzyme is no longer capable of affecting the hydrolysis of ACh; this results in a buildup of the neurotransmitter at a nerve synapse [4]. The toxicity of Coumaran may be attributed to its ability to inhibit the activity of AChE, which catalyses the hydrolysis of the neurotransmitter Ach at nerve synapses and neuromuscular junctions. The heterobicyclics, formate esters, Coumaran and their derivatives are the most neurologically and toxicologically active of the prototype materials [30, 34]. Another probable reason could be that the compounds present in these prototypes are neurologically active compounds that have broad impact across the nervous system which is attenuated by modified acetyl choline and acetate function. These results are in accordance with our earlier work on fumigant toxicity of Coumaran against stored grain insect pests. Therefore, we foresee potential of this biofumigant to be used in organic protection of stored commodities.

CONCLUSIONS

In the present study Coumaran was found to be a potent biofumigant isolated from leaves of L. camara which can be used as a biopesticide. Coumaran acts on the respiratory pigments of the insect and can be used as a biofumigant. The probable reason could be AChE inhibition at the cholinergic synapses of the insects. Further toxicity of Coumaran (if any) to the nontargeted animals is to be studied. Due to insect selectivity and from natural origin Coumaran can be used as a biofumigant against the stored grain pests.

ACKNOWLEDGMENTS

The authors wish to thank the Director, IBSD, for his support and engagement in this study. The authors gratefully acknowledge the financial support by the Department of Biotechnology (DBT), India.

REFERENCES

1. M. D. Lopez and M. J. Pascual-Villalobos, "Mode of inhibition of acetylcholinesterase by monoterpenoids and implications for pest control," Industrial Crops and Products, vol. 31, no. 2, pp. 284–288, 2010.
2. R. D. Fields and G. Burnstock, "Purinergic signalling in neuron-glia interactions," Nature Reviews Neuroscience, vol. 7, no. 6, pp. 423–436, 2006
3. R. Dua and K. D. Gill, "Aluminium phosphide exposure: implications on rat brain lipid peroxidation and antioxidant defence system," Pharmacology & Toxicology, vol. 89, no. 6, pp. 315–319, 2001.
4. R. T. Fukuto, "Mechanism of action of organophosphorus and carbamate insecticides," Environmental Health Perspectives, vol. 87, pp. 245–254, 1990
5. M. G. Lionetto, R. Caricato, A. Calisi, M. E. Giordano, and T. Schettino, "Acetylcholinesterase as a biomarker in environmental and occupational medicine: new insights and future perspectives," BioMed Research International, vol. 2013, Article ID 321213, 8 pages, 2013
6. R. Jbilou, H. Amri, N. Bouayad, N. Ghailani, A. Ennabili, and F. Sayah, "Insecticidal effects of extracts of seven plant species on larval development, -amylase activity and offspring production of Tribolium castaneum (Herbst) (Insecta: Coleoptera: Tenebrionidae)," Bioresource Technology, vol. 99, no. 5, pp. 959–964, 2008
7. Y. Rajashekar, N. Bakthavatsalam, and T. Shivanandappa, "Botanicals as grain protectants," Psyche, vol. 2012, Article ID 646740, 13 pages, 2012.

8. S. B. Markowitz, "Poisoning of an urban family due to misapplication of household organophosphate and carbamate pesticides," Journal of Toxicology: Clinical Toxicology, vol. 30, no. 2, pp. 295–303, 1992.
9. M. K. Nayak, P. J. Collins, J. E. Throne, and J.-J. Wang, "Biology and management of psocids infesting stored products," Annual Review of Entomology, vol. 59, pp. 279–297, 2014
10. J. H. Butler and J. M. Rodriguez, "Methyl bromide in the atmosphere," in The Methyl Bromide Issue, C. H. Bell, N. Price, and B. Chakrabarti, Eds., vol. 1, pp. 27–90, John Wiley & Sons, West Sussex, UK, 1996.
11. M. B. Isman, "Botanical insecticides, deterrents, and repellents in modern agriculture and an increasingly regulated world," Annual Review of Entomology, vol. 51, pp. 45–66, 2006
12. C. L. Cantrell, F. E. Dayan, and S. O. Duke, "Natural products as sources for new pesticides," Journal of Natural Products, vol. 75, no. 6, pp. 1231–1242, 2012
13. C. Lamberth, S. Jeanmart, T. Luksch, and A. Plant, "Current challenges and trends in the discovery of agrochemicals," Science, vol. 341, no. 6147, pp. 742–746, 2013.
14. Y. Rajashekar, L. J. M. Rao, and T. Shivanandappa, "Decaleside: a new class of natural insecticide targeting tarsal gustatory sites," Naturwissenschaften, vol. 99, no. 10, pp. 843–852, 2012
15. Y. Lu, J. Zhong, Z. Wang, F. Liu, and Z. Wan, "Fumigation toxicity of allicin against three stored product pests," Journal of Stored Products Research, vol. 55, pp. 48–54, 2013.
16. G. E. Nenaah, "Potential of using flavonoids, latex and extracts from Calotropis procera (Ait.) as grain protectants against two coleopteran pests of stored rice," Industrial Crops and Products, vol. 45, pp. 327–334, 2013
17. N. Zapata and G. Smagghe, "Repellency and toxicity of essential oils from the leaves and bark of Laurelia sempervirens and Drimys winteri against Tribolium castaneum," Industrial Crops and Products, vol. 32, no. 3, pp. 405–410, 2010
18. P. D. Cox, "Potential for using semiochemicals to protect stored products from insect infestation,"Journal of Stored Products Research, vol. 40, no. 1, pp. 1–25, 2004.

19. Y. Rajashekar, N. Gunasekaran, and T. Shivanandappa, "Insecticidal activity of the root extract ofDecalepis hamiltonii against stored-product insect pests and its application in grain protection," Journal of Food Science and Technology, vol. 47, no. 3, pp. 310–314, 2010
20. S. Miresmailli and M. B. Isman, "Botanical insecticides inspired by plant-herbivore chemical interactions," Trends in Plant Science, vol. 19, no. 1, pp. 29–35, 2014. .
21. M. I. Picollo, A. C. Toloza, G. Mougabure Cueto, J. Zygadlo, and E. Zerba, "Anticholinesterase and pediculicidal activities of monoterpenoids," Fitoterapia, vol. 79, no. 4, pp. 271–278, 2008
22. M. B. Isman, "Plant essential oils for pest and disease management," Crop Protection, vol. 19, no. 8–10, pp. 603–608, 2000
23. I. Ishaaya, A. Barazani, S. Kontsedalov, and A. R. Horowitz, "Insecticides with novel modes of action: mechanism, selectivity and cross-resistance," Entomological Research, vol. 37, no. 3, pp. 148–152, 2007
24. Y. Rajashekar, K. V. Ravindra, and N. Bakthavatsalam, "Leaves of Lantana camara Linn. (Verbenaceae) as a potential insecticide for the management of three species of stored grain insect pests," Journal of Food Science and Technology, 2012
25. Y. Rajashekar, H. Vijaykumar, K. V. Ravindra, and N. Bakthavatsalam, "Isolation and characterization of biofumigant from leaves of Lantana camara for control of stored grain pests," Industrial Crops and Products, vol. 51, pp. 224–228, 2013
26. G. L. Ellman, K. D. Courtney, A. J. Valentino, and R. M. Featherstone, "A new and rapid colorimetric determination of acetylcholinesterase activity," Biochemical Pharmacology, vol. 7, no. 2, pp. 88–95, 1961.
27. O. H. Lowry, N. J. Rosenburg, A. L. Farr, and R. J. Randall, "Protein measurement with the Folin-Phenol reagent," The Journal of Biological Chemistry, vol. 193, no. 1, pp. 265–275, 1951.
28. D. J. Finney, Probit Analysis, Cambridge University Press, Cambridge, UK, 3rd edition, 1971.
29. I. Ishaaya, Biochemical Sites of Insecticide Action and Resistance, Springer, New York, NY, USA, 2000.

30. C. Song and M. E. Scharf, "Neurological disruption by low-molecular-weight compounds from the heterobicyclic and formate ester classes," Pesticide Biochemistry and Physiology, vol. 92, no. 2, pp. 92–100, 2008
31. F. Matsumura, Toxicology of Insecticides, Plenum Press, New York, NY, USA, 2nd edition, 1985.
32. R. Nauen, "Insecticide mode of action: return of the ryanodine receptor," Pest Management Science, vol. 62, no. 8, pp. 690–692, 2006
33. S. J. Yu, the Toxicology and Biochemistry of Insecticides, CRC Press, Boca Raton, Fla, USA, 2008.
34. M. E. Scharf, S. N. Nguyen, and C. Song, "Evaluation of volatile low molecular weight insecticides usingDrosophila melanogaster as a model," Pest Management Science, vol. 62, no. 7, pp. 655–663, 2006

Chapter 3

Pesticides: Environmental Impacts and Management Strategies

Harsimran Kaur Gill[1] and Harsh Garg[2]

[1]University of Florida, Gainesville, FL, USA

[2]The University of Sydney, NSW, Australia

INTRODUCTION

Increase in food production is the prime-most objective of all countries, as world population is expected to grow to nearly 10 billion by 2050. Based on evidence, world population is increasing by an estimated 97 million per year (Saravi and Shokrzadeh, 2011). The Food and Agricultural Organization (FAO) of the United Nations has in-fact issued a sobering forecast that world food production needs to increase by 70%, in order to keep pace with the demand of growing population. However, increase in food production is faced with the ever-growing challenges especially the new area that can be increased for cultivation purposes is very limited (Saravi and Shokrzadeh, 2011). The increasing world population has therefore put a tremendous amount of pressure on the existing agricultural system so that food needs can be met from the same current resources like land, water etc. In the process of increasing crop production, herbicides, insecticides, fungicides, nematicides, fertilizers and soil amendments are now being used in higher quantities than in the past. These chemicals have mainly come into the picture since the introduction of synthetic insecticides in 1940, when organochlorine (OCl) insecticides were first used for pest management. Before this introduction, most weeds, pests, insects and diseases were controlled using sustainable practices such as cultural, mechanical, and physical control strategies.

Pesticides have now become an integral part of our modern life and are used to protect agricultural land, stored grain, flower gardens as well as to eradicate the pests transmitting dangerous infectious diseases. It has been estimated that globally nearly $38 billion are spent on pesticides each year (Pan-Germany, 2012). Manufacturers and researchers are designing new formulations of pesticides to meet the global demand. Ideally, the applied pesticides should only be toxic to the target organisms, should be biodegradable and eco-friendly to some extent (Rosell et al., 2008). Unfortunately, this is rarely the case as most of the pesticides are non-specific and may kill the organisms that are harmless or useful to the ecosystem. In general, it has been estimated that only about 0.1% of the pesticides reach the target organisms and the remaining bulk contaminates the surrounding environment (Carriger et al., 2006). The repeated use of persistent and non-biodegradable pesticides has polluted various components of water, air and soil ecosystem. Pesticides have also entered into the food chain and have bioaccumulated in the higher tropic level. More recently, several human acute and chronic illnesses have been associated with pesticides exposure (Mostafalou and Abdollahi, 2012). Below, we have detailed the effect of pesticides on target and non-target organisms including earthworms, predators, pollinators, humans, fishes, amphibians, and birds. Additionally, impact of pesticides on soil, water and air ecosystems is also discussed. Furthermore, an eco-friendly practice (Integrated Pest Management (IPM) approach) has been detailed as a strategy that could minimize the use of pesticides.

EFFECTS OF PESTICIDES ON TARGET ORGANISMS

Over the past era there has been an increase in the development of pesticides to target a broad spectrum of pests. The increased quantity and frequency of pesticide applications have posed a major challenge to the targeted pests causing them to either disperse to new environment and/or adapt to the novel conditions (Meyers and Bull, 2002; Cothran et al., 2013). The adaptation of the pest to the new environment could be attributed to the several mechanisms such as gene mutation, change in population growth rates, and increase in number of generations etc. This has ultimately resulted in increased incidence of pest resurgence and appearance of pest species that are resistant to pesticides.

Pesticide Resistance

"Resistance may be defined as a heritable change in the sensitivity of a pest population that is reflected in the repeated failure of a product to achieve the expected level of control when used according to the label recommendation for that pest species" (IRAC, 2013). Resistant individuals tend to be rare in a normal population, but indiscriminate use of chemicals can eliminate normal susceptible populations and thereby providing the resistant individuals a selective advantage in the presence of a pesticide. Resistant individuals continue to multiply in the absence of competition and eventually become the dominant portion of the population over generations. As majority of the individuals of a population are resistant, the insecticide is no longer effective thus causing the appearance or development of insecticide resistance.

Resistance is the most serious bottleneck in the successful use of pesticides these days. The intensive use of pesticides has led to the development of resistance in many targeted pest species around the globe (Tabashnik et al., 2009). Number of resistant insects and mite species had risen to 600 by the end of 1990, and increased to over 700 by the end of 2001. This trend is likely to be continued in 21st century as well. Resistance has been found in different insecticides groups e.g., 291 species have developed cyclodiene resistance, followed by DDT (263 species), organophosphates (260 species), carbamates (85 species), pyrethroids (48 species), fumigants (12 species), and other (40 species) (Dhaliwal et al., 2006). Important crop pests, parasites of livestock, common urban pests and disease vectors in some cases have developed resistance to such an extent that their control has become exceedingly challenging (Van Leeuwen et al., 2010; Gondhalekar et al., 2011). However, many factors such as genetics, biology/ecology and control operations influence the development of pesticide resistance (Georghiou and Taylor, 1977).

Insecticide bioassays using whole insects continue to be one of the most widely used approaches for detecting resistance (Brown and Brogdon, 1987; Gondhalekar et al., 2013) despite some associated drawbacks. In the past two decades, however, several new methods employing advanced biochemical and molecular techniques, and combination of insecticide bioassays have been developed for detecting insecticide resistance (Symondson and Hemingway, 1997; Scharf et al., 1999; Zhou et al., 2002). Some examples of these techniques are enzyme electrophoresis, enzyme assays, immuno-assays, allele-specific polymerase chain reaction (PCR) etc.

Pest Resurgence

Pest resurgence is defined as the rapid reappearance of a pest population in injurious numbers following pesticide application. Use of persistent and broad spectrum pesticides that kills the beneficial natural enemies is thought to be the leading cause of pest resurgence. However, resurgence is known to occur due to several reasons, for example, increase in feeding and reproductive rates of insect pests, due to application of sub-lethal doses of pesticides, and sometimes elimination of a primary pest provides favorable conditions for the secondary pests to become primary/key pests (Dhaliwal et al., 2006). There are many pesticide-induced pest outbreaks reported in walnut (*Juglans regia*) (Bartlett and Ewart, 1951), hemlock (*Conium maculatum*) (McClure, 1977), soybeans (*Glycine max*) (Shepard et al., 1977), and cotton (*Gossypium hirsutum*) (Bottrell and Rummel, 1978). Among these, brown plant hopper (BPH) (*Nilaparvata lugens* (Stal)) in rice (*Oryza sativa* L.) cultivation has gained a major importance in Asian countries (Chelliah and Heinrichs, 1984). In general, natural BPH populations were kept under check by natural enemies including mirid bugs, ladybird beetles, spiders and various pathogens. However, pesticides have not only destroyed the natural enemies (Fabellar and Heinrichs, 1986), but have influenced the fecundity of BPH females (Wang et al., 2010) further enhancing their resurgence. Additionally, the resurgence of bed bug, *Cimex lectularius* (Davies et al., 2012) and cotton bollworm*Helicoverpa armigera* (Mironidis et al., 2013) have been reported due to insecticide resistance and indiscriminate use of pesticides.

EFFECTS OF PESTICIDES ON NON-TARGET ORGANISMS

The effect of pesticides on non-target organisms has been a source of worldwide attention and concern for decades. Adverse effects of applied pesticides on non-target arthropods have been widely reported (Ware, 1980). Unfortunately, natural insect enemies e.g., parasitoids and predators are most susceptible to insecticides and are severely affected (Aveling, 1977; Vickerman, 1988). The destruction of natural enemies can exacerbate pest problems as they play an important role in regulating pest population levels. Usually, if natural enemies are absent, additional insecticide sprays are required to control the target pest. In some cases, natural enemies that normally keep minor pests under check are also affected and this can result in secondary pest outbreaks. Along with natural enemies, population of soil arthropods is also drastically disturbed because of indiscriminate pesticide application in agricultural systems. Soil invertebrates including nematodes, springtails, mites, micro-arthropods, earthworms, spiders, insects and other small organisms make up the soil food web and enable decomposition of organic compounds such as leaves, manure, plant residues etc. They are essential for the maintenance of soil structure, transformation and mineralization of organic matter. Pesticide effects on above mentioned soil arthropods therefore negatively impact several links in the food web. The following are the examples of non-target organisms that are adversely impacted by pesticides.

Earthworms

Earthworms represent the greatest proportion of terrestrial invertebrates (>80%) (Yasmin and D'Souza, 2010) and play a significant role in improving soil fertility by decomposing the organic matter into humus. Earthworms also play a major role in improving and maintaining soil structure, by creating channels in soil that enable the process of soil aeration and drainage. However, their diversity, density and biomass are strongly influenced by soil management. They are considered as an important indicator of soil quality in agricultural ecosystems (Paoletti, 1999). Earthworms are affected by various agricultural practices and indiscriminate use of pesticides is one of the leading practices affecting them (Pelosi et al., 2013).

Pesticide applications can cause decline in earthworm populations. For example, carbamate insecticides are very toxic to earthworms and some organophosphates have been shown to reduce earthworm populations (Edwards, 1987). Similarly, a field study conducted in South Africa has also reported that earthworms were influenced detrimentally due to chronic and intermittent exposures to chlorpyrifos and azinphos methyl, respectively (Reinecke and Reinecke, 2007). Various scientific studies reported that pesticides influence earthworm growth, reproduction (cocoon production, number of hatchlings per cocoon, and incubation period) in a dose-dependent manner (Yasmin and D'Souza, 2010). Earthworms exposed to different kind of pesticides showed rupturing of cuticle, oozing out of coelomic fluid, swelling, and paling of body that led to softening of body tissues (Solaimalai et al., 2004). Similarly a study carried out in France showed that the combination of insecticides and fungicides at different concentrations caused neurotoxic effects in earthworms (Schreck et al., 2008). Increased exposure period and higher dose of insecticides can also cause physiological damage (cellular dysfunction and protein catabolism) to earthworms (Schreck et al., 2008).

Predators

Predators are organisms that live by preying on other organisms and they play a very crucial role in keeping pest populations under control. Predators (beneficial organisms) are also an important part of the "biological control" approach which is one component of the integrated pest management strategy discussed later. In some of the examples cited below, pesticides were the main cause for decline in predator population:

- In brinjal (*Solanum melongena* L.) ecosystem, spraying with cypermethrin and imidacloprid caused higher mortality of coccinellids, braconid wasps and predatory spiders compared to when sprayed with bio-pesticides and neem (*Azadirachta indica*) based insecticides (Ghananand et al., 2011).
- Species diversity, richness and evenness of collembola, and numbers of spiders were found to be lower in chlorpyrifos treated plots compared with control, in grassland pastures in UK (Fountain et al., 2007).
- Studies were carried out to investigate the effects of chemicals on soil arthropods in agricultural area near Everglades National Park, USA. It was found that higher number of arthropods (including predators such as coccinelids and spiders) were present in non-sprayed fields compared to fields sprayed with insecticides and herbicides (Amalin et al., 2009).
- In foliar application, all the systemic neonicotinoids such as imidacloprid, clothianidin, admire, thiamethoxam and acetamiprid were found highly toxic to natural enemies in comparison with spirotetramat, buprofezin and fipronil (Kumar et al., 2012).

Additionally, pesticides can also affect predator behavior and their life-history parameters including growth rate, development time and other reproductive functions. For example, in the eastern USA, glyphosate-based herbicides affected behavior and survival of spiders and ground beetles, apart from affecting arthropod community dynamics that can also influence biological control in an agroecosystem (Evans et al., 2010). Similarly, dimethoate was shown to significantly decrease the body size, haemocyte counts and reduction of morphometric parameters on carabid beetle (*Pterostichus melas italicus*), in Calabria, Italy (Giglio et al., 2011).

Pollinators

Pollinators are biotic agents that play a very important role in pollination process. Some of the recognized pollinators are different species of bees, bumble bees (*Bombus* spp.), honey bees (*Apis* spp.), fruit flies, some beetles, and birds (e.g., hummingbirds, honeyeaters, and sunbirds etc.). Pollinators can be used as bioindicators of ecosystemic processes (process by which physical, chemical, biological events help connecting organisms with their environment) in many ways as their activities are affected by environmental stress caused by parasites, competitors, diseases, predators, pesticides and habitat modifications (Kevan, 1999). However, using pesticides causes direct loss of insect pollinators and indirect loss to crops because of the lack of adequate populations of pollinators (Fishel, 2011).

Pesticide application also affects various activities of pollinators including foraging behaviour, colony mortality and pollen collecting efficiency. Most of our current knowledge about effects of pesticides on change in pollinator behaviour has come from various bee studies as they comprise 80% of the insect pollinator population. For instance, many laboratory studies have demonstrated the lethal and sub-lethal effects of neonicotinoid insecticides (imidacloprid, acetamiprid, clothianidin, thiamethoxam, thiacloprid, dinotefuran and nitenpyram) on foraging behavior, learning and memory abilities of bees (Blacquie`re et al., 2012). Worker bee (female bees that lack full reproductive capacity and play many other roles in bee colony) mortality, decreased pollen collecting efficiency and eventually colony collapse occur due to pesticides (neonicotinoid and pyrethroid) application (Gill et al., 2012). In addition to this, non-lethal exposure of honey bees to neonicotinoid insecticide (thiamethoxam) causes high mortality due to homing failure at a level that could put a risk of colony collapse (Henry et al., 2012). Sub-lethal doses of imidacloprid (the most commonly used pesticide worldwide) affected longevity and foraging in honey bees (*A. mellifera*). *Nosema ceranae* (*Nosema* invades the intestinal tracts of adult bees causing colony collapse disorder (CCD) and nosema disease/nosemosis, which consequently lead to decrease in honey production). Microsporidial infections increased significantly in gut of bees from imidacloprid treated hives. It has been anticipated that interactions between pathogens and imidacloprid pesticide could be a main reason for worldwide honey bee colony mortality, including

CCD (Pettis et al., 2012; Wu et al., 2012). There are also reports that imidacloprid reduced brood production due to decline in the fecundity of bumble bees (*B. terrestris*) (Laycock et al., 2012; Whitehorn et al., 2012). On the other hand, little work has been done on the impact of pesticides on wild pollinators. For example, a field study carried out in Italy on an agricultural field found lower bumblebee and butterfly species richness associated with pesticide application. They also found that bees (insect pollinators) were at higher risk from pesticide use (Brittain et al., 2010).

Humans

The deleterious effects of pesticides on human health have started to grow due to their toxicity and persistence in environment and ability to enter into the food chain. Pesticides can enter the human body by direct contact with chemicals, through food especially fruits and vegetables, contaminated water or polluted air. Both acute and chronic diseases can result from pesticide exposure and these are summarized below:

Acute Illness

Acute illness generally appears a short time after contact or exposure to the pesticide. Pesticide drift from agricultural fields, exposure to pesticides during application and intentional or unintentional poisoning generally leads to the acute illness in humans (Dawson et al., 2010; Lee et al., 2011b). Several symptoms such as headaches, body aches, skin rashes, poor concentration, nausea, dizziness, impaired vision, cramps, panic attacks and in severe cases coma and death could occur due to pesticide poisoning (Pan-Germany, 2012). The severity of these risks is normally associated with toxicity and quantity of the agents used, mode of action, mode of application, length and frequency of contact with pesticides and person that is exposed during application (Richter, 2002). About 3 million cases are reported worldwide every year that occur due to acute pesticides poisoning. Out of these 3 million pesticide poisoning cases, 2 million are suicide attempts and the rest of these are occupational or accidental poisoning cases (Singh and Mandal, 2013). Suicide attempts due to acute pesticide poisoning are mainly the result of widespread availability of pesticides in rural areas (Richter, 2002; Dawson et al., 2010). Several strategies have been proposed to reduce the incidences that occur due to acute pesticide poisoning such as restricting the availability of pesticides, substituting the pesticide with a less toxic but with an equally effective alternative and by promoting use of personal protection equipment (Murray and Taylor, 2000; Konradsen et al., 2003). Strict laws regulating pesticide sales along with preventive health programs and community development efforts are needed to enforce such strategies.

Chronic Illness

Continued exposure to sub-lethal quantities of pesticides for a prolonged period of time (years to decades), results in chronic illness in humans (Pan-Germany, 2012). Symptoms are not immediately apparent and manifest at a later stage. Agricultural workers are at a higher risk to get affected, however general population is also affected especially due to contaminated food and water or pesticides drift from the fields (Pan-Germany, 2012). Incidences of chronic diseases have started to grow as pesticides have become an increasing part of our ecosystem. There is mounting evidence that establish a link between pesticides exposure and the incidences of human chronic diseases affecting nervous, reproductive, renal, cardiovascular, and respiratory systems (Mostafalou and Abdollahi, 2012). The list of chronic diseases that are linked to prolonged pesticide exposure by various studies is summarized in Table 1.

Table 1: The List of chronic diseases that are linked to the exposure to pesticides

Diseases	**References**
Cancer (Childhood and adult brain cancer; Renal cell cancer; lymphocytic leukaemia (CLL); Prostate Cancer)	Lee et al., 2005; Shim et al., 2009; Heck et al., 2010; Xu et al., 2010; Band et al., 2011; Cocco et al., 2013
Neuro degenerative diseases including Parkinson disease, Alzheimer disease	Elbaz et al., 2009; Hayden et al., 2010;Tanner et al., 2011
Cardio-vascular disease including artery disease	Abdullah et al., 2011; Andersen et al., 2012
Diabetes (Type 2 Diabetes)	Son et al., 2010; Lee et al., 2011a
Reproductive disorders	Petrelli and Mantovani, 2002; Greenlee et al., 2003
Birth defects	Winchester et al., 2009; Mesnage et al., 2010

Hormonal imbalances including infertility and breast pain	Xavier et al., 2004
Respiratory diseases (Asthma, Chronic obstructive pulmonary disease (COPD))	Chakraborty et al., 2009; Hoppin et al., 2009

Several mechanisms have been illustrated that link development of chronic diseases with pesticide exposure. Direct interaction of pesticides with genetic material resulting in DNA damages and chromosomal aberration is considered to be one of the primary mechanisms that lead to the chronic diseases such as cancer etc (Mostafalou and Abdollahi, 2012). In this context, several studies report an increase in frequency of chromosomal aberration, sister chromatid exchange, and breakage in DNA strand in pesticide applicators who worked in agricultural fields (Grover et al., 2003; Santovito et al., 2012). Similar to this, pesticides are also known to induce epigenetic changes (heritable changes without any alteration in DNA sequences) through DNA methylation, histone modifications and expression of non-coding RNAs. For example, neurotoxic pesticide paraquat has been implicated to induce the Parkinson's disease (PD) through epigenetic changes by promoting histone acetylation (Song et al., 2010). Pesticides may also induce oxidative stress by increasing reactive oxygen species (ROS) through altering levels of antioxidant enzymes such as superoxide dismutase, glutathione reductase and catalase (Agrawal and Sharma, 2010). Several health problems such as Parkinson disease, disruption of glucose homeostasis have been linked with pesticides induced oxidative stress (Mostafalou and Abdollahi, 2012).

PESTICIDES AND SOIL ENVIRONMENT

A major fraction of the pesticides that are used for agriculture and other purposes accumulates in the soil. The indiscriminate and repeated use of pesticides further aggravates this soil accumulation problem. Several factors such as soil properties and soil micro-flora determine the fate of applied pesticides, owing to which it undergoes a variety of degradation, transport, and adsorption/desorption processes (Weber et al., 2004; Laabs et al., 2007; Hussain et al., 2009). The degraded pesticides interact with the soil and with its indigenous microorganisms, thus altering its microbial diversity, biochemical reactions and enzymatic activity (Hussain et al., 2009; Munoz-Leoz et al., 2011). A summary of the effects of pesticides on its various components are given below:

1. Pesticides that reach the soil can alter the soil microbial diversity and microbial biomass. Any alteration in the activities of soil microorganisms due to applied pesticides eventually leads to the disturbance in soil ecosystem and loss of soil fertility (Handa et al., 1999). Numerous studies have been undertaken which highlight these adverse impacts of pesticides on soil microorganisms and soil respiration (Dutta et al., 2010; Sofo et al., 2012). In addition to this, exogenous applications of pesticides could also influence the function of beneficial root-colonizing microbes such as bacteria and arbuscular mycorrhiza (AM), fungi and algae in soil by influencing their growth, colonization and metabolic activities etc (Debenest et al., 2010; Menendez et al., 2010; Tien and Chen, 2012).

The pesticides that reach the soil can interact with soil microflora in several ways:

- It can adversely affect the growth, microbial diversity or microbial biomass of the soil microflora. For example, sulfonylurea herbicides- metsulfuron methyl, chlorsulfuron and thifensulfuron methyl were reported to reduce the growth of the fluorescent bacteria *Pseudomonas* strains that were isolated from an agricultural soil (Boldt and Jacobson, 1998). The *Pseudomonas* spp. is known to play an important ecological role in the soil habitat (Boldt and Jacobson, 1998), and hence its reduction can adversely affect soil fertility. Similarly, benomyl, captan and chlorothalonil were reported to suppress the peak soil respiration (an indicator of microbial biomass) in an unamended soil by 30–50% (Chen et al., 2001b).
- Pesticide application may also inhibit or kill certain group of microorganisms and outnumber other groups by releasing them from the competition (Hussain et al., 2009). For example, increase in bacterial biomass by 76% was reported in response to endosulfan application and that reduced the fungal biomass by 47% (Xie et al., 2011).
- Applied pesticide may also act as a source of energy to some of the microbial group which may lead to increase in their growth

and disturbances in the soil ecosystem. For example, bacterial isolates collected from wastewater irrigated agricultural soil showed the capability to utilize chlorpyriphos as a carbon source for their growth (Bhagobaty and Malik, 2010).

- Pesticides can alter and/or reduce the functional structure and functional diversity of microorganisms, but increase the microbial biomass (Lupwayi et al., 2009). In contrast, application of pesticides can also reduce the microbial biomass while increasing the functional diversity of microbial community. For example, methamidophos and urea decreased the microbial biomass and increased the functional diversity of soil as determined by microbial biomass and community level physiological profiles (Wang et al., 2006).

1. Pesticides may also adversely affect the soils vital biochemical reactions including nitrogen fixation, nitrification, and ammonification by activating/deactivating specific soil microorganisms and/or enzymes (Hussain et al., 2009; Munoz-Leoz et al., 2011). The synergistic and additive interactions between pesticides, micro-organisms and soil properties ultimately govern increase or decrease in rate of soil biochemical reactions. For example, populations of the *Azospirillum* spp. bacteria and the rate of ammonification was reported to increase at a particular pesticide concentration (i.e 2.5 to 5.0 kg ha^{-1}) in both laterite and vertisol soils planted to groundnut (*Arachis hypogaea* L.). But the tested pesticides exerted antagonistic interactions on the population of *Azospirillum* spp. and ammonification at higher concentrations (7.5 and 10.0 kg ha^{-1}) (Srinivasulu et al., 2012a).
2. Pesticides have also been reported to influence mineralization of soil organic matter, which is a key soil property that determines the soil quality and productivity. For example, a significant reduction in soil organic matter was found after the application of four herbicides (atrazine, primeextra, paraquat, and glyphosate) (Sebiomo et al., 2011). However, soil organic matter then increased after continuous application from the second to the sixth week of herbicide treatment.
3. Pesticides that reach the soil may also disturb local metabolism or can alter the soil enzymatic activity (Gonod et al., 2006; Floch et al., 2011). Soil in general contains an enzymatic pool which

comprises of free enzymes, immobilized extracellular enzymes and enzymes excreted by (or within) microorganisms that are indicator of biological equilibrium including soil fertility and quality (Mayanglambam et al., 2005; Hussain et al., 2009). Degradation of both pesticides and natural substances in soil is catalyzed by this enzymatic pool (Floch et al., 2011; Kizilkaya et al., 2012). Due to this, measuring the change in enzymatic activity has now been classified as a biological indicator to identify the impact of chemical substances including pesticides on soil biological functions (Garcia et al., 1997; Romero et al., 2010). In fact, it has generally been assumed that measuring the change in enzyme activity is an earlier indicator of soil degradation as compared to the chemical or physical parameters (Dick et al., 1994). Several studies have already been undertaken which indicate both increase and decrease in activities of soil enzymes such as hydrolases, oxidoreductases, and dehydrogenase (Ismail et al., 1998; Megharaj et al., 1999). A description of pesticides interactions with soil enzymes has been summarized in Table 2.

Table 2: A summary of the effects of pesticides on different soil enzymes

Enzyme (Function in soil)	**Examples of the pesticides applied**	**Comments**
Nitrogenase (An enzyme used by organisms to fix atmospheric nitrogen gas).	Carbendazim, Imazetapir, Thiram, Captan, 2,4-D, Quinalphos, Monocrotophos, Endosulfan, -HCH, Butachlors	Pesticide reduced or inhibited the nitrogenase activity in laboratory or field conditions (Chalam et al., 1996; Martinez-Toledo et al., 1998; Niewiadomska, 2004; Niewiadomska and Klama, 2005; Prasad et al., 2011)/Pesticides stimulated the nitrogenase activity (Patnaik et al., 1995)
Phosphatase (hydrolyzes organic P compounds to inorganic P)	2,4-D, Nitrapyrin, Monocrotophos, Chlorpyrifos, Mancozeb and Carbendazim	Inhibited (Tu, 1981); Activity increased, but higher concentration or increasing incubation period has inhibitory effects (Madhuri and Rangaswamy, 2002; Srinivasulu et al., 2012b)
Urease (catalyzes the hydrolysis of urea into CO_2 and NH_3 and is a key component in the nitrogen cycle in soils)	Isoproturon, Benomyl, Captan, Diazinon, Profenofos	Increase in urease activity (Chen et al., 2001a; Nowak et al., 2004), Pesticide reduced/inhibited urease activity (Abdel-Mallek et al., 1994; Ingram et al., 2005)

Dehydrogenase (DHA): (an oxidoreductase enzyme that catalyzes the removal of hydrogen)	Azadirachtin, Acetamiprid, Quinalphos,Glyphosate	Positive/stimulatory influence on the DHA (Singh and Kumar, 2008; Kizilkaya et al., 2012)/Initially inhibited but later on activity was restored (Andrea et al., 2000; Mayanglambam et al., 2005)
Invertase (hydrolyzes sucrose to fructose and glucose)	Atrazine, Carbaryl, Paraquat	Inhibited invertase activity (Gianfreda et al., 1995; Sannino and Gianfreda, 2001)
β-glucosidase (hydrolyzes disaccharides in soil to form -glucose)	Metalaxyl, Ridomil gold plus copper	Enzyme activity increased and then decreased (Sukul, 2006) or inhibited (Demanou et al., 2004)
Cellulase (hydrolyzes cellulose to D-glucose)	Benlate, Captan, Brominal	Inhibited enzyme activity (Arinze and Yubedee, 2000; Omar and Abdel-Sater, 2001)
Arylsulphatase (an enzyme that hydrolyzes aryl sulfates)	Cinosulfuron, Prosulfuron, Thifensulfuron methyl, Triasulfuron	Decreased enzyme activity (Sofo et al., 2012)

Several environmental factors control the bioavailability, degradation and effect of pesticides on soil microorganisms in addition to the persistence, concentration and toxicity of the applied pesticides. These include soil texture, presence of organic matter, vegetation and cultural practices (Murage et al., 2007). For instance, a mixture of compost and straw was found to have the capability of bio-degrading different mixtures of fungicides that are usually applied in vineyards when tested under laboratory conditions (Coppola et al., 2011). Similarly, persistence of the herbicide imazapyr was reported to be different in three Argentinean soils (Tandil, Anguil, and Cerro Azul sites) and its half-life was negatively associated with soil pH, iron and aluminum content, and positively related with clay content (Gianelli et al., 2013). Additionally, level of soil moisture is also one of the most important factors that regulates pesticide bioavailability and degradation, as water acts as solvent for pesticide movement and diffusion, and is essential for microbial functioning (Pal and Tah, 2012). For example, degradation of herbicide saflufenacil was found to be faster at field capacity for Nada, Crowley and Gilbert soils as compared to the saturated soil conditions (Camargo et al., 2013).

It is important to monitor the response of soil microbial communities and various enzymatic activities to pesticide exposure in order to reduce their deleterious effects. A combination of both cultivation-dependent (e.g., community-level physiological profiling (CLPP), measuring overall rates of microbial activity) and cultivation-independent (e.g., DNA sequence information, proteomics of environmental samples) methods can be applied to measure and interpret the effects of pesticide exposure (Imfeld and Vuilleumier, 2012). With the advent of efficient new sequencing techniques and metagenomics, the scope of deploying cultivation independent methods for measuring bacterial diversity and function in soil ecosystem has been further increased. Metagenomics approach has been applied already to measure microbial diversity for a range of soil systems including contaminated sites (Ono et al., 2007) and land managed with different cultural practices (Souza et al., 2013). Such high-tech approaches hold the key for future methods to measure the mode of adaptation ecosystem to different pesticides and in development of new methods to better manage pesticide applications.

A careful screening of pesticide effects on soil microflora should be done in laboratory before their field applications. This is because pesticides tend to accumulate in soil due to repeated applications over time and can pose adverse effects on soil microflora even though they are applied at recommended doses (Ahemad et al., 2009). For instance, Ahemad and Khan (2011) reported the highest toxicity to plant growth promoting characteristics of the *Bradyrhi zobium* sp. when its strain MRM6 was grown with three times the recommended field rates of glyphosate, imidacloprid and hexaconazole. Similarly, Dunfield et al. (2000) assessed the effects of the fungicides captan and thiram at rates of 0.25-2 g a.i. kg^{-1}on the survival and phenotypic characteristics of bacteria *Rhizobium leguminosarum* bv. *viceae*, strain C1. They found that even though both captan and thiram significantly reduced the numbers of rhizobia recovered from seed and altered the FAME (fatty acid methyl ester) and biological profiles of recovered rhizobia, it was only the highest concentrations of captan that affected nodulation and plant growth. Similarly, herbicide mesotrione affected soil microbial communities, but the effects were only detected at doses far exceeding the recommended field rates (Crouzet et al., 2010). Overall, it is crucial to comprehend the role of pesticides in perturbing soil environment, so that the risk of pesticide contamination and its consequent adverse impacts on soil environment can be evaluated.

PESTICIDES IN WATER AND AIR ECOSYSTEM

Pesticide residues in water are a major concern as they pose a serious threat to biological communities including humans. There are different ways by which pesticides can get into water such as accidental spillage, industrial effluent, surface run off and transport from pesticide treated soils, washing of spray equipments after spray operation, drift into ponds, lakes, streams and river water, aerial spray to control water-inhibiting pests (Carter and Heather, 1995; Singh and Mandal, 2013). Pesticides generally move from fields to various water reservoirs by runoff or in drainage induced by rain or irrigation (Larson et al., 2010). Similarly, the presence of pesticides in air can be caused by number of factors including spray drift, volatilization from the treated surfaces, and aerial application of pesticides. Extent of drift depends on: droplet size and wind speed. The rate of volatilization is dependent on time after pesticide treatment, the surface on which the pesticide settles, the ambient temperature, humidity and wind speed and the vapor pressure of the ingredients (Kips, 1985). The volatility or semi-volatility nature of the pesticide compounds similarly constitutes an important risk of atmospheric pollution of large cities (Trajkovska et al., 2009). For instance, organophosphorus (OP) pesticides were identified from environmental samples of air and surface following agricultural spray applications in California and Washington (USA) (Armstrong et al., 2013). In Italian forests, indiscriminate use of pesticides and its active metabolites has led to the contamination of water bodies and ambient air, possibly affecting the health of aquatic biota fishes, amphibians and birds (Trevisan et al., 1993). The following section describes the effect of pesticides on fishes, amphibians and birds.

Fishes

Fishes are an important part of marine ecosystem as they interact closely with physical, biological and chemical environment. Fishes provide food source for other animals such as sea birds and marine mammals and thus fishes form an integral part of the marine food web. A lot of research has been carried out to examine the impact of pesticides on decline in fish population (Scholz et al., 2012). Pesticides have been directly linked to causing fish mortality worldwide. For example, 27 freshwater fish species are found to be affected by "plant protection products" (PPP) in Europe (Ibrahim et al., 2013). Another pesticide pentachlorophenol (NaPCP) is reported to cause large numbers of fish mortality in the rice fields of Surinam (Vermeer et al., 1970). Pesticides not only impact the fish but also food webs related to them. The persistent pesticides (organochlorine pesticides and polychlorinated biphenyls) have already been found in the major Arctic Ocean food webs (Hargrave et al., 1992). A survey was conducted to examine the influence of pesticides on aquatic community in West Bengal, India. Many body tissues of the fish such as gills, alimentary canal, liver and brain of carp and catfish were found drastically damaged by pesticides. It was reported that such level of pesticides in fish could harm the fish consumers as well (Konar, 2011).

Several examples are available where pesticides impacted the vital fish organs and behavior. Organophosphate pesticide "Abate" has the potential to alter the vitellogenesis (the process whereby yolky eggs are produced) of catfish (*Heteropneustes fossilis* (Bloch.)), which can severely affect catfish farming (Kumari, 2012). Another major effect of toxic contaminants is on olfaction in fishes since it can affect activities such as mating, locating food, avoiding predators, discriminating kin and homing etc (Tierney et al., 2010). Simultaneous exposure of trematode parasite (*Telogaster opisthorchis*), freshwater fish (*Galaxias anomalus*) and snails to high glyphosate concentrations significantly reduced their survival and development. Within 24 hrs of exposure to higher glyphosate concentrations, 100% mortality of individuals was found (Kelly et al., 2010).

The impact of pesticides within an aquatic environment is influenced by their water solubility and uptake ability within an organism (Pereira et al., 2013). For example, Clomazone, a popular herbicide, is particularly water soluble; a property that increases its likelihood of contaminating surface and groundwater. The hydrophilic (water-loving) or lipophobic (fat-hating) nature of this pesticide makes it less available in the fatty tissues of an organism (Pereira et al., 2013). Further to this, the toxicity of chemical (e.g., endosulfan in this case) in juvenile rainbow trout (*Oncorhynchus mykiss*) was affected by alkalinity, temperature of water and size of the fish (Capkin et al., 2006).

Pesticides in natural water within the acceptable concentration range can still pose harmful effects. Kock-Schulmeyer et al. (2012) found that even if the pesticide levels found in Llobregat River basin of Spain were within the European Union Environmental Quality Standards, they still accounted for a low to high ecotoxicological risk for aquatic organisms, especially algae and macro-invertebrates. Proper measures should be taken while disposing of expired pesticides, so that their discharge into the water bodies does not danger the aquatic life. This is because the alteration in water pH by expired insecticides can lead to acute toxicity of different fish (Satyavani et al., 2011).

Amphibians

Amphibians are ectothermic, tetrapod vertebrates of class Amphibia. They inhabit a wide variety of habitats, with most species living within fossorial, arboreal, terrestrial, and freshwater aquatic ecosystems. The global decline in the amphibian population has become an environmental concern worldwide. Many amphibian species are on the brink of extinction with 7.4% listed as critically endangered, and at least 43.2% experiencing some sort of population decrease (Stuart et al., 2004). There could be multitude of reasons for decline in amphibian species diversity, but pesticides appear to be playing an important role. Global warming and climate change are leading to more variable and warmer temperatures which may have increased the impact of pesticides on amphibian populations (Relyea, 2003; Johnson et al., 2013).

Many studies showed that amphibians are susceptible to environmental contaminants due to their permeable skin, dual aquatic-terrestrial cycle and relatively rudimentary immune system (Kerby et al., 2010). Several studies showing the impact of pesticides on amphibians are being mentioned here. It has been reported that the world's most commonly used herbicide (Roundup (Glyphosate)) may have far reaching effects on non-target amphibians (Relyea, 2012). Roundup, a globally used herbicide caused high mortality of larval tadpoles (3 different species in North America) and juvenile frogs under natural conditions in an outdoor pond mesocosm (Relyea, 2005a). Most of the evidence supported the toxic effects of pesticides on juvenile European common frogs (*Rana temporaria*) in an agricultural field that was over sprayed. Mortality of frogs ranged from 100% after 1 hour to 40% after 7 days at the recommended concentrations of pesticides

(Bruhl et al., 2013). It was found that population of the wood frog (*Lithobates sylyaticus*) near an agricultural area was more resistant to common insecticide (chlorpyrifos), but not to the common herbicide (Roundup). However, no evidence was reported that resistance carried a performance cost when facing competition and the fear of predation (Cothran et al., 2013).

Further to this, pesticides indirectly affect amphibian populations by influencing growth of aquatic communities such as fungi, zooplankton, and phytoplankton as they are one of their prime energy resources. Malathion is the most commonly used broad-spectrum insecticide in United States. It is legal to spray malathion over aquatic habitats to control mosquitoes (Family: Culicidae), that vector malaria and West Nile Virus. A study found that even low concentration of malathion caused direct and indirect effects on aquatic communities (Relyea, 2012). For example, indirect effect of malathion led to decrease in zooplankton diversity, that led to increase in phytoplankton, a decrease in periphyton, and finally decrease in growth of frog tadpoles (Relyea and Hoverman, 2008). Moreover, it was found that repeated applications of low doses had largest impacts than single high dose application of malathion on an aquatic system (Relyea and Diecks, 2008). A comprehensive study was conducted to examine the effect of globally used pesticides including insecticides (carbaryl, malathion, and herbicides (glyphosate, 2, 4-D)) on aquatic communities (algae, 25 animal species). Species richness reduced differentially, 15% with carbaryl, 30% with malathion, and 22% with roundup, whereas 0% with 2, 4-D. It was found that Roundup completely eliminated two species of tadpoles and led to 70% decline in tadpole species (Relyea, 2005b). Another study demonstrated that frogs (*Rana pipiens*) living in agricultural area, where they experienced higher exposure to chemicals were smaller in size and weight than frogs living in area exposed to low-levels of chemicals. It suggests that frogs living in agricultural areas might have more vulnerability to infections and diseases due to their smaller size and alternation in their immune system (decrease in number of splenocytes and phagocytic activity) (Christin et al., 2013).

Birds

Birds are a diverse group, and apart from their distinct songs and calls, showy displays and bright colors adding enjoyment to lives of humans, they play a very critical role in food chains and webs in our ecosystems. Birds are also called "aerial acrobats" consuming different kinds of insects such as mosquitoes, European corn borer moth (*Ostrinia nubilalis*), Japanese beetles (*Popillia japonica*), and many other insect species that are considered as some of the most serious agricultural and health pests. Birds are important biotic components of an ecosystem and help in maintaining a natural equilibrium of insect populations by predating on them. In absence of birds, outbreaks of insect pest populations would become more common, ultimately leading to increased pesticide use. Pesticides exposure by different means such as direct ingestion of pesticide granules and treated seeds, treated crops, direct exposure to sprays, contaminated water, or feeding on contaminated prey, and baits cause birds mortality (Fishel, 2011; Guerrero et al., 2012). In USA, almost 50 pesticides are known for killing song birds, game birds, seabirds, shorebirds, and raptors (BLI, 2004).

Pesticides have a potential to alter behavior and reproduction of birds. Some of the examples cited here, using different synthetic chemicals including carbamates, organochlorines, and organophosphates can cause a decline in the populations of raptorial birds by altering their feeding behavior and reproduction (Mitra et al., 2011). A large area in the world is under rice and therefore cultivation and volume of pesticides applied in rice field is quite significant. Many different kinds of organochlorines, cholinesterase-inhibiting insecticides including carbofuran, monocrotophos, phorate, diazinon, fenthion, phosphamidon, methyl parathion and azinphos-methyl along with fungicides, herbicides and molluscicides are being used in rice fields. Some of these chemicals are highly toxic to birds causing mortality and some chemicals even have the potential to affect their reproductive systems (Parsons et al., 2010). Indirect effects of pesticides, through food chain have been proposed as a possible factor in decline of farmland bird species. Insecticides applied in breeding season can affect breeding performance of corn bunting (*Miliaria calandra*) and yellowhammer (*Emberiza citrinella*) (Boatman et al., 2004).

Pesticides, especially insecticides such as carbamates and organophosphates have the potential to cause bird mortality due to their high toxicity (Hunter, 1995). Further to this, insecticides and fungicides pose a most prominent threat to ground-nesting farmland birds as compared to other agricultural practices. The decline of

US grassland birds is attributed to acute pesticide toxicity and not agricultural intensification as previously thought (Mineau and Whiteside, 2013). An estimate suggests that 672 million birds are directly exposed to pesticides every year on farmlands, and 10% of these birds die due to acute toxic effects of pesticides (Williams, 1997). A study was conducted in rice fields of Surinam to examine the effects of pesticides, pentachlorophenol (NaPCP) on birds. NaPCP was sprayed for the purpose of killing*Pomacea* snails. Large numbers of dead sick/dead egrets, herons and jacana birds were found during the period of pesticide application. Pentachlorophenol and endrin levels in these birds suggested that ingestion of contaminated food was the probable cause of sickness and mortality (Vermeer et al., 1970).

PESTICIDES AND BIOMAGNIFICATION

The increase in concentration of pesticides due to its persistent and non-biodegradable nature in the tissues of organisms at each successive level of food chain is known as biomagnification. Due to this phenomenon, organisms at the higher levels of food chain experience greater harm as compared to those at lower levels. Several studies have been undertaken that demonstrate enhanced amount of toxic compounds with increase in trophic levels. For example, out of 36 species collected from three lakes of northeastern Louisiana (USA) that were found to contain residues of 13 organochlorines, tertiary consumers such as green-backed heron (*Butorides striatus*), and snakes etc., contained the highest residues as compared to secondary consumers (bluegill (*Lepomis macrochirus*), blacktail shiner (*Notopis venustus*)) (Niethammer et al., 1984). Similarly, significantly higher concentrations of dichlorodiphenyltrichloroethane (4,4′-DDE) were found in the top consumer fish in Lake Ziway, catflish (*Clarias gariepinus*) than in lower consumers, Nile tilapia (*Oreochromis niloticus*), tilapia (*Tilapia zillii*) and goldfish (*Carassius auratus*) (Deribe et al., 2013). Some of the adverse effects of pesticides on non-target organisms such as fish, amphibians and humans discussed in the above section have also occurred as a result of biomagnifications of the toxic compounds. For example, reproductive failure and population decline in the fish-eating birds (e.g., gulls, terns, herons etc.) was observed as a result of

DDE induced eggshell thinning (Grasman et al., 1998). The extent of biomagnifications increases with increase in persistence and lipophilic (fat-loving) characteristics of the particular pesticide. As a result of this, organochlorines are known to have higher biomaginification rate and are more persistent in a wider range of organisms as compared to organophsphates (Favari et al., 2002). It is important to do the risk assessments associated with the pesticides on the basis of their bioaccumulation and biomagnifications before considering them for agricultural purposes.

STRATEGIES FOR PESTICIDE MANAGEMENT

There are a relatively few pesticide resistance management tactics that have been proposed risk-free and have a reasonable chance of success under a variety of different circumstances. Headmost among these are: monitoring of pest population in field before any pesticide application, alteration of pesticides with different modes of action, restricting number of applications over time and space, creating or exploiting refugia, avoiding unnecessary persistence, targeting pesticide applications against the most vulnerable stages of pest life cycle, using synergists which can enhance the toxicity of given pesticides by inhibiting the detoxification mechanisms. The most difficult challenge in managing resistance is not the unavailability of appropriate methods but ensuring their adoption by growers and pest control operators (Denholm et al., 1998; Dhaliwal et al., 2006).

Pest resurgence is a dose-dependent process and there are ways to tackle this problem using correct dosage of effective and recommended pesticides. Resurgence problem occurs due to a number of reasons. One of them is due to farmers' tendency to apply low-dose insecticides due to economic constraints that lead to inadequate and ineffective control of pests. Pest resurgence also occurs due to reduced biological control (most common with insects), reduced competition (most common with weeds; monocots vs. dicots), direct stimulation of pest (due to sub-lethal dose), and improved crop growth.

In the current scenario, optimized use of pesticides is important to reduce environmental contamination while increasing their effectiveness against target pest. This way we can reduce pesticide resistance as well as pest resurgence problems. This has led to the consideration of rational use of pesticides, and the physiological and ecological selectivity of pesticides. Physiological selectivity is characterized by differential toxicity between taxa for a given insecticide. However, ecological selectivity refers to the modification of operational procedure in order to reduce unnecessary destruction to non-target organisms (Dent, 2000). Farmers should focus to use insecticides that are more toxic to target species than their natural enemies which could help to reduce resurgence to some extent (Dhaliwal et al., 2006).

One should consider adopting an Integrated Pest Management (IPM) approach for controlling pests, as these practices are designed to have minimal environment disturbance. The aim of IPM is not only to reduce indiscriminate pesticide use but also to substitute hazardous chemicals with safe chemistries. IPM is a process of achieving long-term, environmentally safe pest control using wide variety of technology and other potential pest management practices. According to National Academy of Science, "IPM refers to an ecological approach in pest management in which all available necessary techniques are consolidated in a unified program so that populations can be managed in such a manner that economic damage is avoided and adverse side effects are minimized" (NAS, 1969). In European arable systems, applied multi-disciplinary research and farmer incentives to encourage the adoption of innovative IPM strategies are essential for development of sustainable maize-based cropping systems. These IPM strategies can contribute immensely to address the European strategic commitment to the environmentally sustainable use of pesticides (Vasileiadis et al., 2011). The added cost and time to do an IPM approach is sometimes a difficult task for growers, but government and extension services can help in convincing and encouraging growers to go for IPM strategy for eco-friendly and long term pest control. We have already discussed earlier that continuous use of pesticides leads to pesticide resistance and pest resurgence problem. To avoid these issues we can always go for other potential management options that include cultural and physical control, host plant resistance, biocontrol, and the use of biopesticides etc.

Cultural Control

Historically, cultural control methods were the farmer's most important tool of preventing crop losses. Cultural control for pest management has been adopted by growers throughout the world for a long time due to its environmentally friendly nature and minimal costs (Gill et al., 2013). Cultural control practices are regular farm operations, which are used to destroy the pests or to prevent them from causing plant damage. Several methods of cultural control have been practiced, such as crop rotation, sanitation, soil solarization, timed planting and harvest, use of resistant varieties, certified seeds, allelopathy, intercropping or "companion planting", use of farmyard manure, and living and organic mulches (Altieri et al., 1978; Dent, 2000; Dhaliwal et al., 2006). Soil solarization (McSorley and Gill, 2010; Gill and McSorley, 2011b) and organic mulches (Gill and McSorley, 2011a) alone and their integration (Gill and McSorley, 2010) were reported as economical and eco-friendly technique for controlling soil-surface arthropods (various insects, and nematodes) (Gill et al., 2010; Gill et al., 2011) and weeds (Gill et al., 2009; Gill and McSorley, 2011b). More effective cultural control can be achieved by synchronizing existing practices with life cycles of pests. This way the weakest link in their life cycle is subjected to adverse climatic conditions.

Large insect populations are killed automatically by farmers when they expose them to adverse climatic conditions through agricultural practices like weeding, ploughing, and hoeing. Ploughing of agricultural field allows turnover of the upper layer of soil while burying the weeds and residues from last year. For example, in South Africa, about 70% of overwintering populations of spotted stalk borer (*Chilo partellus*) and maize stalk borer (*Busseola fusca*) in grain sorghum (*Sorghum bicolor* L.) and maize (*Zea mays* L.) fields were destroyed by slashing the plants. Ploughing and discing of plant residues after slashing further destroyed 24% population on grain sorghum and 19% on maize (Kfir, 1990). Planting dates (Goyal and Kanta, 2005a), and barrier crops (teosinte (*Zea* spp.) and pearl millet (*Pennisetum glaucum* (L.)) (Goyal and Kanta, 2005b) were found to be effective against maize stem borer (*Chilo partellus*) in India. The brown seaweeds *Spatoglossum asperum* and *Sargassum swartzii* can be used as manure to protect plants (tomato (*Solanum lycopersicum* L.) in this case) from root rotting fungi, (*Macrophomina phaseolina, Rhizoctonia solani* and *Fusarium solani*) and root-knot nematode (*Meloidogyne javanica*) and for providing necessary nutrients to plants (Sultana et al., 2012). In India, rodents are pests in agriculture, horticulture, forestry, animal husbandry as well as in human dwellings and rural and urban storage facilities. Cultural methods, such as clean cultivation, proper soil tillage and crop scheduling, barriers, repellents and proofing that reduce the rodent harbourage, food sources and immigration may have long lasting effects (Parshad, 1999).

Physical and Mechanical Control

Managing pest populations using devices which affect them physically or alter their physical environment is called physical control. Exposure to sun rays, steaming, moisture management especially for stored grain pests, and light traps for attracting various kinds of moths, beetles and other pests are different methods used in physical control. For example steaming woolen winter clothes help in eliminating population of the woolly bear moth, *Antherenus vorax* (waterhouse) (Dhaliwal et al., 2006). Hot water treatment of plant storage products like corns, and bulbs helps to kill many concealed pests such as eelworms and bulb flies. Superheating of empty grain storage godowns to a temperature of 50°C for 10-12 hours helps killing hibernating stored grain pests. Exposure of cotton seeds to sun's heat in thin layers for 2-3 days during summer helps in killing diapausing larvae of pink bollworm (*Pectinophora gossypiella* Saunders) (Dhaliwal et al., 2006).

Mechanical control refers to suppression of pest population by manual devices. It includes various practices such as hand picking, trapping and suction devices, clipping, pruning and crushing of infested shoots and floral parts, and exclusion by screens and barriers to keep away house flies (*Musca domestica*), mosquitoes and other pests. In south-eastern Australia, the common starling (*Sturnus vulgaris*) is an established invasive avian pest that is now making incursions into Western Australia which is currently free of this species. Trapping with live-lure birds is suggested to be the most cost-effective and widely implemented starling control technique (Campbell et al., 2012). Numerous wildlife species such as coyotes (*Canis latrans* Say), squirrels (Sciuridae family), and birds are known pests of California agriculture in the United States. For these pests, different non-lethal control options including habitat modification, exclusionary devices, and baiting are generally preferred (Baldwin et al., 2013). Mechanical weed control is mainly associated with tillage practices which are performed with special tools such as harrows, hoes, and brushes in growing crops. Increased knowledge about side effects of herbicides has further driven the interest in adoption of mechanical weed control thus increasing the prevalence of organic farming (Rueda-Ayala et al., 2010; Jat et al., 2011). Trapping using yellow colored sticky traps is an effective way for controlling tephritid flies (Dhaliwal et al., 2006).

Host Plant Resistance

Host plant resistance (HPR) is the genetic ability of the plant to improve its survival and reproduction by a range of adaptations as compared to the other cultivars when exposed to the same level of pest infestation. HPR offers the most effective, economical and eco-friendly method of pest control (Sharma and Ortiz, 2002), and is considered to be a key element of the IPM strategy. Due to this, identifying and developing HPR has always been a major thrust area of plant breeding, and a number of breeding programs aiming to develop pest resistant crops have been deployed in almost all the cultivated crop species. For example, identification and/or development of resistant varieties in maize against European corn borer (*Ostrinia nubilalis* (Hubner)) (Dhaliwal et al., 2006) , brassica against cabbage butterfly (*Pieris brassicae* Linn.) (Chahil and Kular, 2013), wheat (*Triticum aestivum* L.) and rye (*Secale cereale*L.) against *Fusarium* diseases (Miedaner, 1997) *Brassica* sp. against Sclerotinia disease (Garg et al., 2008), and in rice against bacterial blight (Khush et al., 1989). Additionally, availability and access to various germplasm collections have increased the scope of widening the gene-pool of cultivated crops and further identifying and developing HPR. Wild species are especially known to possess a rich repository of genes against various defense traits as they have evolved under different geographic locations. Considerable progress has been made where identification and/or transfer of resistance gene from wild to cultivated species against various pest species has been achieved such as in potato (*Solanum tuberosum*L.) against late blight (*Phytophthora infestans*) from ten wild *Solanum sp.* (Colon and Budding, 1988), wheat against powdery mildew (*Erysiphe graminis*) from wild emmer wheat (*Triticum dicoccoides*) (Reader and Miller,

1991) and mustard (*Brassica juncea*) against *Sclerotinia sclerotiorum* from*Erucastrum cardaminoides* (Garg et al., 2010).

Use of Biotechnology and Molecular Approaches for Developing Resistant Genotype

The advent of new biotechnological and molecular approaches has opened the way to develop resistant genotype that could not only reduce the pesticides application, but it also has a potential to be a part of IPM. Development of resistant genotypes in classical breeding is met with several challenges such as it is time consuming, desired traits are linked with the undesirable traits (linkage drag) and most importantly lack of resistant genotypes in the gene pool. On the other hand, use of biotechnology in crop improvement ensures the development of pest-resistant genotypes in a comparatively short period of time and minimizes the effects of linkage drag. One of the classic examples where biotechnology was successfully deployed to develop resistant genotype is by the synthesis of transgenic plants which involves modifying plant traits by inserting foreign DNA from a different species (De la Pena et al., 1987). A number of different crops including cotton, rice, mustard, and maize have been modified up to now to engineer the genotypes against various biotic stresses (Ahmad et al., 2012). One of the most successful examples of synthesis of transgenic genotype against pest resistance is in cotton where the gene coding for Bt toxin from the bacterium *Bacillus thuringiensis* (Bt) was inserted leading cotton genotypes to produce Bt toxin in its tissue (Pray et al., 2002; Wu et al., 2008). The lepidopteran larvae that fed on the transgenic plants were killed due to Bt toxin eventually decreasing the amount of pesticide applied to the field. Examples of transgenic crops that have been developed with a potential to reduce pesticides use are abound and few of them include potato lines against potato tuber moth (*Phthorimaea operculella*) expressing Cry1Ab (Kumar et al., 2010), rice against yellow stem borer (*Scirpophaga incertulas*) expressing potato proteinase inhibitor 2 (Bhutani et al., 2006) and oilseed rape lines resistant to various fungal attack over-expressing tomato chitinase gene (Grison et al., 1996).

Another strategy where biotechnology and molecular approaches have been deployed to combat biotic stresses involves the use of RNA interference (RNAi) technique. This technique primarily uses transgenic plants expressing double stranded RNA (dsRNA) and that reduces the messenger RNA (mRNA) levels (with a high specificity and fidelity) of a crucial gene in the target pest upon feeding (Price and Gatehouse, 2008; Kos et al., 2009). This ultimately interferes with the development and survival of the target pest. RNAi has emerged as a powerful functional genomics approach and it has been used to engineer several crops against number of insect-pests. For example, RNAi technique was used in tobacco genotype that targeted the gene "integrase splicing factor" in root knot nematode, *Meloidogyne incognita* nematode eventually leading to the decrease in the number of nematodes 6-7 weeks post inoculation (Yadav et al., 2006). When such an advanced and effective approach is combined with IPM, it has a great potential to decrease chemical use in agricultural and other ecosystems.

Biological Control

The process of using natural enemies of particular pests to reduce their populations to such a level where economic losses are either eliminated or suppressed is called biological control. Traditionally the most important biocontrol agents are parasitoids, predators and pathogens. Biological control involves three major techniques, *viz.*, introduction, conservation, and augmentation of natural enemies. Biocontrol agents include vertebrates, nemathelminthes (flatworms, and roundworms), arthropods (spiders, mites, and insects), pathogens like viruses, bacteria, protozoa, fungi and rickettsiae all of which play a dynamic role in natural regulation of insect and mite populations (Dhaliwal et al., 2006). In 1762, the Indian Mynah, *Acridotheres tristis* (Linnaeus), was introduced to control red locust in Mauritius. First significant success in controlling a pest was achieved on the suggestion of C. V. Riley of California (USA) in 1888. The Vedalia beetle (*Rodolia cardinalis* (Mulsant)), was introduced from Australia into California (USA) for the control of cottony-cushion scale (*Icerya purchasi* maskell) on citrus plants. This scale insect had been accidentally introduced earlier from Australia (Dhaliwal et al., 2006).

Biological control of weeds has been very successful worldwide. There are about 41 species of weeds which have been successfully controlled using insects and pathogens as biocontrol agents. Also, 3 weed species have been controlled using native fungi as mycoherbicides (Mcfadyen, 2000). A total of 12 insects were released in Australia against prickly pear (*Opuntia stricta*), out of these, *Dactylopius opuntiae* and*Cactoblastis cactorum* were responsible for the successful control of prickly pear weed (Julien and Griffiths, 1998). In the past decade, Australia has released 43 species of arthropods and pathogens in 19 different projects for successful biological control of many exotic weeds. Effective biological control was achieved in several projects and outstanding success was achieved in the control of rubber vine (*Cryptostegia grandiflora*), and bridal creeper (*Asparagus asparagoides*) (Palmer et al., 2010).

Examples of biological control are available for other organisms like helminthes, nematodes, fungi, bacteria etc. A nematophagous fungus (*Monacrosporium thaumasium*) was found to be effective in controlling cyathostomin, one of the most important helminthes in tropical region of southeastern Brazil (Tavela et al., 2011). *Trichoderma* species are free-living fungi that have been used to control a broad range of plant pathogenic fungi, viruses, bacteria and nematodes especially root-knot nematodes (*Meloidogyne javanica* and *M. incognita*) (Sharon et al., 2011).

Biorational Pesticides

Biorational pesticides/ biopesticides are considered as third-generation pesticides that are rapidly gaining popularity. The word biorational is derived from two words, "biological" and "rational", which means pesticides of natural origin that have limited or no adverse effects on the environment or beneficial organisms. Biopesticides encompass a broad array of microbial pesticides, plant pesticides and biochemical pesticides which are derived from micro-organisms and other natural sources, and processes involving the genetic incorporation of DNA into agricultural commodities. The most commonly used biopesticides include biofungicides (e.g., *Trichoderma* spp.), bioherbicides (*Phytopthora* spp.), bioinsecticides (spore forming bacteria, *Bacillus thuringiensis*, and *B. popilliae*, Actinomycetes), naturally occurring fungi (*Beauveria bassiana*), microscopic roundworms (Entomopathogenic nematodes), Spinosad, insect hormones and insect growth regulators (Gupta and Dikshit, 2010; Singh et al., 2013).

Applications of microbial insecticide, *Chromobacterium subtsugae* for suppression of pecan weevil (*Curculio caryae* (Horn)), and combination of eucalyptus extract and microbial insecticide, *Isaria fumosorosea* (Wize) for control of black pecan aphid (*Melanocallis caryaefoliae* (Davis)) were found promising as alternative insecticides (Shapiro-Ilan et al., 2013). Entomopathogenic nematodes (EPNs) belonging to the families Heterorhabditidae and Steinernematidae are potentially used in South Africa as biocontrol agents against vine mealybug (*Planococcus ficus* (Signoret)) (le Vieux and Malan, 2013). Spinosad was found effective in controlling Colorado potato beetle (*Leptinotarsa decemlineata*) in Iran, and is recommended for use in IPM program for Colorado potato beetle (Soltani and Agricultural, 2011). In China, entomopathogenic fungus (*Beauveria bassiana*) has shown great potential for the management of some bark beetle species including red turpentine beetle (RTB) (*Dendroctonus valens* LeConte), a destructive invasive pest (Zhang et al., 2011).

The allelopathic properties of plants can be exploited successfully as a tool for weed and pathogen reduction. In a rice field, application of allelopathic plant material @ 1-2 tonne/ha reduced weed diversity by 70% and increased yield by 20%. Numerous growth inhibitors identified from these allelopathic plants are responsible for their allelopathic properties and may be a useful source for the future development of bio-herbicides and pesticides (Xuan et al., 2005). A combination of coleopteran-active toxin, *Bacillus thuringiensis* Cry3Aa protoxin and protease inhibitors, especially a potato carboxypeptidase inhibitor, have efficiency in preventing damage to stored products and grains by stored grain coleopteran pests (Oppert et al., 2011).

Chemical Control

Sometimes cultural and other agro-technical practices are not sufficient to keep pest population below economic injury level (lowest pest population density that will cause economic crop damage). Therefore, the chemical control agents are resorted to both as preventive and curative measures to minimize the insect pest damage. A good pesticide should be potent against pests, should not endanger the health of humans and non-target organisms, and should ultimately break down into harmless compounds so that it does not persist in environment. Both relative and specific toxicities of the pesticide need to be estimated in order to determine its potency.

It is very important to know spray droplet size and density chemical dosage, application timing, which can provide adequate pest control. There is also a need for research into the development of suitable packaging and disposal procedures, as well as refining of the application equipment. All of these shall rationalize the use of pesticides, so that they can be used in an acceptable way.

Very strict laws should be enacted to protect wildlife and other non-target organisms. Following directions on the pesticide label can prevent injury to non-target organisms. However, when these directions are not followed, benefits from pesticides can be outweighed by the harm and risk associated with pesticides (Fishel, 2011). During pesticide application, things that need to be considered are timing of insecticide application, dosage and persistence, and selective placement of insecticides as discussed below.

Timing of Pesticide Application

The timing of pesticide application is an important factor to consider before doing any pesticide application. Appropriate application time can ensure not only maximum impact on the target organisms but also least impact on beneficial organisms. Pesticide application timing mainly depends on availability of weather window, time at which pests can be best controlled, and when least damage will be caused to non-target organisms and environment. Flowering period in crops and middle of the day are the times when bees are more prone to insecticides. Hence, insecticide application should not happen at those times to avoid decline in bee populations. Time of insecticide application should coincide with the most vulnerable stage of insect life cycle. Monitoring of insects in the field is thus extremely important for knowing the stage of insect pest in the field. Monitoring systems are available for most of the insect pests, but spray regime or experiments need to be carried out to determine the most appropriate time for insecticide application for insects for which monitoring systems are not available (Hull and Starner, 1983;Richter and Fuxa, 1984).

Time of the day and season of the year are also important to consider when making pesticide applications. The early morning and evening hours are often the best times for pesticide application because windy conditions are more likely to occur around midday when the temperature warms near the ground level. This causes hot air to rise quickly and mix rapidly with the cooler air above it, favoring drift. During stable conditions, a layer of warm air can stay overhead and not promote mixing with colder air that stays below and closer to the ground. Inversions tend to dissipate during the middle of the day when wind currents mix the air layers. It is very important that applicators recognize thermal inversions and do not spray under those conditions. A temperature or thermal inversion is a condition that occurs naturally and exists when the air at ground level is cooler than the temperature of the air above it. Wind speed is the most important weather factor influencing drift. High wind speeds will move droplets downwind and deposit them off the target. On the other hand, dead calm conditions are never recommended due to likelihood of temperature inversions (Fishel and Ferrell, 2013). Drifting of pesticides increases the possibility of injury to pollinators, humans, domestic animals and wildlife. It is recommended not to spray in wind speed above 2.5 miles/second which otherwise can cause excessive drift and eventually contamination of adjacent areas (Matthews, 1981). Pesticide application should not be made just before rain because pesticides can be washed off by the rain without any impact on the target pest.

Dosage and Persistence

Pesticide dose should be sufficient but no greater than the level required for best results. The pesticide manufacturer sets the dose to ensure an acceptable level of control, producing acceptable residue levels, and maximizing returns per unit of formulated insecticide. Persistent pesticides have their benefit of longer persistence on the target and therefore requires less frequent spraying compared to non- persistent pesticides. But care should be taken while using persistent pesticides since these might diminish benefits from natural enemies even at lower doses. If an insecticide is persistent in nature, chances of insecticide residues being harmful to natural enemies are greatly increased (Dent, 2000).

Selective Placement

Distribution of pesticides in the field should be such that maximum target cover is achieved. Usually only about 1% of the applied pesticides is able to reach its target, while a large amount of it is wasted. Understanding the pest biology and behavior is critical as it can provide information on pest's habitat, fecundity, feeding etc., which can be important considerations before applying pesticides. Most of the pesticides are applied in liquid form and thus the droplet size is very important in determining their effectiveness. Small droplets provide better coverage and greater likelihood of coming in contact with the target compared to larger droplets that can bounce off the plant surface very easily. The disadvantage with smaller and bigger droplets is the increased chance of drift and therefore a balance has to be considered between smaller droplets to obtain the maximum effectiveness and reduced drift.

In situations where crops are grown on beds covered with plastic mulch, pesticides should be injected into soil at the time the plastic is laid or injected afterward through drip irrigation system to achieve maximum pesticide effectiveness. For termite (Order: Isoptera) treatments, sometimes perimeter application of insecticides is required around structures/buildings. Additionally, liquids that form foams following injections can be injected into small spaces that are or might be inhabited by termites or other small creatures.

CONCLUSIONS

Although, pesticides were used initially to benefit human life through increase in agricultural productivity and by controlling infectious disease, their adverse effects have overweighed the benefits associated with their use. The above discussion clearly highlights the severe consequences of indiscriminate pesticide use on different environmental components. Some of the adverse effects associated with pesticide application have emerged in the form of increase in resistant pest population, decline in on beneficial organisms such as predators, pollinators and earthworms, change in soil microbial diversity, and contamination of water and air ecosystem. The persistent nature of pesticides has impacted our ecosystem to such an extent that pesticides have entered into various food chains and into the higher trophic levels such as that of humans and other large mammals. Some of the acute and chronic human illnesses have now emerged as a consequence of intake of polluted water, air or food.

This is the time that necessitates the proper use of pesticides to protect our environment and eventually health hazards associated with it. Alternative pest control strategies such as IPM that deploys a combination of different control measures such as cultural control, use of resistant genotype, physical and mechanical control, and rational use of pesticide could reduce the number and amount of pesticide applications. Further, advanced approaches such as biotechnology and nanotechnology could facilitate in developing resistant genotype or pesticides with fewer adverse effects. Community development and various extension programs that could educate and encourage farmers to adopt the innovative IPM strategies hold the key to reduce the deleterious impact of pesticides on our environment.

ACKNOWLEDGEMENTS

We would like to say special thanks to Drs. Gaurav Goyal (Territory Agronomist, Monsanto), Ameya D. Gondhalekar (Research Assistant Professor, Purdue University), Siddharth Tiwari (Entomologist, BASF), and Matthew R. Tarver (Research Entomologist, USDA) for their valuable suggestions and comments for improving this manuscript.

REFERENCES

1. Abdel-Mallek AY, Moharram AM, Abdel-Kader MI, Omar SA. Effect of soil treatment with the organophosphorus insecticide profenofos on the fungal flora and some microbial activities. Microbiological Research 1994;149:167-171.
2. Abdullah NZ, Ishaka A, Samsuddin N, Mohd Rus R, Mohamed AH. Chronic organophosphate pesticide exposure and coronary artery disease: finding a bridge. IIUM Research, Invention and Innovation Exhibition (IRIIE) 2011;223.

3. Agrawal A, Sharma B. Pesticides induced oxidative stress in mammalian systems: a review. International Journal of Biological and Medical Research (IJBMR) 2010;1:90-104.
4. Ahemad M, Khan MS, Zaidi A, Wani PA. Remediation of herbicides contaminated soil using microbes. In: Khan MS, Zaidi A, Musarrat J (eds.) Microbes in sustainable agriculture. Nova Science Publishers Inc., New York;2009. p261-284.
5. Ahemad M, Khan MS. Effect of pesticides on plant growth promoting traits of greengram-symbiont, *Bradyrhizobium* sp. strain MRM6. Bulletin of Environmental Contamination and Toxicology 2011;86:384-388.
6. Ahmad P, Ashraf M, Younis M, Hu X, Kumar A, Akram NA, Al-Qurainy F. Role of transgenic plants in agriculture and biopharming. Biotechnology Advances 2012;30:524-540.
7. Altieri MA, Francis CA, Schoonhoven AV, Doll JD. A review of insect prevalence in maize (*Zea mays* L.) and bean (*Phaseolus vulgaris* L.) polycultural systems. Field Crops Research 1978;1:33-49.
8. Amalin DM, Peña JE, Duncan R, Leavengood J, Koptur S. Effects of pesticides on the arthropod community in the agricultural areas near the Everglades National Park. Proceedings of Florida State Horticultural Society 2009;122:429-437.
9. Andersen HR, Wohlfahrt-Veje C, Dalgård C, Christiansen L, Main KM, Nellemann C, Murata K, Jensen TK, Skakkebaek NE, Grandjean P. Paraoxonase 1 polymorphism and prenatal pesticide exposure associated with adverse cardiovascular risk profiles at school age. PloS One 2012;7:e36830.
10. Andrea MM, Peres TB, Luchini LC, Pettinelli Junior A. Impact of long-term pesticide application on some soil biological parameters. Journal of Environmental Science and Health B 2000;35:297-307.
11. Arinze AE, Yubedee AG. Effect of fungicides on *Fusarium* grain rot and enzyme production in maize (*Zea mays* L.). Global Journal of Pure and Applied Sciences 2000;6:629-634.
12. Armstrong JL, Fenske RA, Yost MG, Galvin K, Tchong-French M, Yu J. Presence of organophosphorus pesticide oxygen analogs in air samples. Atmospheric Environment 2013;66:145-150.

13. Aveling C. The biology of Anthocorids (Heterophera: Anthocoridae) and their role in the integrated control of the damson-hop aphid (*Phorodon humili* Schrank). PhD Thesis, University of London;1977.

14. Baldwin RA, Salmon TP, Schmidt RH, Timm RM. Wildlife pests of California agriculture: regional variability and subsequent impacts on management. Crop Protection 2013;46:29-37.

15. Band PR, Abanto Z, Bert J, Lang B, Fang R, Gallagher RP, Le ND. Prostate cancer risk and exposure to pesticides in British Columbia farmers. Prostate 2011;71:168-183.

16. Bartlett B, Ewart WH. Effect of parathion on parasites of *Coccus hesparidum*. Journal of Economic Entomology 1951;44:344-347.

17. Bhagobaty RK, Malik A. Utilization of chlorpyrifos as a sole source of carbon by bacteria isolated from wastewater irrigated agricultural soils in an industrial area of western Uttar Pradesh, India. Research Journal of Microbiology 2010;3:293-307.

18. Bhutani S, Kumar R, Chauhan R, Singh R, Choudhury VK, Choudhury JB. Development of transgenic indica rice plants containing potato proteinase inhibitor 2 gene with improved defense against yellow stem borer. Physiological and Molecular Biology of Plants 2006;12:43-52.

19. Blacquiere T, Smagghe G, Cornelis AM, Gestel van, Mommaerts V. Neonicotinoids in bees: a review on concentrations, side-effects and risk assessment. Ecotoxicology 212;21:973-992.

20. BLI. State of the world's birds. Birdlife International 2004. (http://www/biodiversityinfo.org/default.php.).

21. Boatman ND, Brickle NW, Hart JD, Milsom TP, Morris AJ, Murray AWA, Murray KA, Robertson PA. Evidence for the indirect effects of pesticides on farmland birds. Ibis 2004;146:131-143.

22. Boldt TS, Jacobsen CS. Different toxic effects of the sulphonylurea herbicides metsulfuron methyl, chlorsulfuron and thifensulfuron methyl on fluorescent pseudomonads isolated from an agricultural soil. FEMS Microbiology Letters 1998;161:29-35.

23. Bottrell DG, Rummel DR. Response of *Heliothis* populations to insecticides applied in an area-wide reproduction diapause boll weevil suppression program. Journal of Economic Entomology 1978;71:87-92.

24. Brittain CA, Vighi M, Bommarco R, Settele J, Potts SG. Impacts of a pesticide on pollinator species richness at different spatial scales. Basic and Applied Ecology 2010;11:106-115.

25. Brown TM, Brogdon WG. Improved detection of insecticides resistance through conventional and molecular techniques. Annual Review of Entomology 1987;32:145-162.

26. Bruhl CA, Schmidt T, Pieper S, Alscher A. Terrestrial pesticide exposure of amphibians: an underestimated cause of global decline? Scientific Reports 2013;3. (Article number 1135). DOI:10.1038/srep01135.

27. Camargo ER, Senseman SA, Haney RL, Guice JB, McCauley GN. Soil residue analysis and degradation of saflufenacil as affected by moisture content and soil characteristics. Pest Management Science 2013;69:1291-1297. DOI:10.1002/ps.3494.

28. Campbell S, Cook S, Mortimer L, Palmer G, Sinclair R, Woolnough AP. To catch a starling: testing the effectiveness of different trap and lure types. Wildlife Research 2012;39:183-191.

29. Capkin E, Altinok I, Karahan S. Water quality and fish size affect toxicity of endosulfan, an organochlorine pesticide, to rainbow trout. Chemosphere 2006;64:1793-1800.

30. Carriger JF, Rand GM, Gardinali PR, Perry WB, Tompkins MS, Fernandez AM. Pesticides of potential ecological concern in sediment from South Florida canals: an ecological risk prioritization for aquatic arthropods. Soil and Sediment Contamination 2006;15:21-45.

31. Carter AD, Heather AIJ. Pesticides in groundwater: In: Best GA, Ruthven AD. (eds.) Pesticides- developments, impacts, and controls. The Royals Society of Chemistry, London, UK;1995. p113-123.

32. Chahil GS, Kular JS. Biology of *Pieris brassicae* (Linn.) on different *brassica* species in the plains of Punjab. Journal of Plant Protection Research 2013;53:53-59.

33. Chakraborty S, Mukherjee S, Roychoudhury S, Siddique S, Lahiri T, Ray MR. Chronic exposures to cholinesterase-inhibiting pesticides adversely affect respiratory health of agricultural workers in India. Journal of Occupational Health 2009;51:488-497.

34. Chalam AV, Sasikala C, Ramana CV, Rao PR. Effect of pesticides on hydrogen metabolism of *Rhodobacter sphaeroides* and *Rhodopseudomonas palustris*. FEMS Microbiology Ecology 1996;19:1-4.
35. Chelliah S, Heinrichs EA. Factors contributing to brown planthopper resurgence. Resurgence of the brown planthopper, *Nilaparvata lugens* (Stål) following insecticide application. Paper presented at the 9th Annual Conference of the Pest Control Council of the Philippines, Philippine International Convention Center, Manila, May 3-6;1984. p36 (mimeo).
36. Chen SK, Edwards CA, Subler S. A microcosm approach for evaluating the effects of the fungicides benomyl and captan on soil ecological processes and plant growth. Applied Soil Ecology 2001a;18:69-82.
37. Chen SK, Edwards CA, Subler S. Effect of fungicides benomyl, captan and chlorothalonil on soil microbial activity and nitrogen dynamics in laboratory incubations. Soil Biology and Biochemistry 2001b;33:1971-1980.
38. Christin MS, Menard L, Giroux I, Marcogliese DJ, Ruby S, Cyr D, Fournier M, Brousseau P. Effects of agricultural pesticides on the health of *Rana pipiens* frogs sampled from the field. Environmental Science and Pollution Research 2013;20:601-611.
39. Cocco P, Satta G, Dubois S, Pili C, Pilleri M, Zucca M, Mannetje AM, Becker N, Benavente Y, Sanjose SDe, Foretova L, Staines A, Maynadie M, Nieters A, Brennan P, Miligi L, Ennas MG, Boffetta P. Lymphoma risk and occupational exposure to pesticides: results of the Epilymph study. Occupational and Environmental Medicine 2013;70:91-98.
40. Colon LT, Budding DJ. Resistance to late blight (*Phytophthora infestans*) in ten wild *Solanum* species. Euphytica 1988;39:77-86.
41. Coppola L, Comitini F, Casucci C, Milanovic V, Monaci E, Marinozzi M, Taccari, M, Ciani M, Vischetti, C. Fungicides degradation in an organic biomixture: impact on microbial diversity. New Biotechnology 2011;29:99-106.
42. Cothran RD, Brown JM, Relyea RA. Proximity to agriculture is correlated with pesticide tolerance: evidence for the evolution

of amphibian resistance to modern pesticides. Evolutionary Applications 2013;6:832-841.

43. Crouzet O, Batisson I, Besse-Hoggan P, Bonnemoy F, Bardot C, Poly F, Bohatier J, Mallet C. Response of soil microbial communities to the herbicide mesotrione: a dose-effect microcosm approach. Soil Biology and Biochemistry 2010;42:193-202.
44. Davies TGE, Field LM, Williamson MS. The re-emergence of the bed bug as a nuisance pest: implications of resistance to the pyrethroid insecticides. Medical and Veterinary Entomology 2012;26:241-254.
45. Dawson AH, Eddleston M, Senarathna L, Mohamed F, Gawarammana I, Bowe SJ, Manuweera G, Buckley NA. Acute human lethal toxicity of agricultural pesticides: a prospective cohort study. PLoS Medicine 2010;7:e1000357.
46. De la Pena A, Lörz H, Schell J. Transgenic rye plants obtained by injecting DNA into young floral tillers. Nature 1987;325:274-276.
47. Debenest T, Silvestre J, Coste M, Pinelli E. Effects of pesticides on freshwater diatoms. In: Reviews of environmental contamination and toxicology, Springer, New York 2010;203:87-103.
48. Demanou J, Monkiedje A, Njine T, Foto SM, Nola M, Serges H, Togouet Z, Kemka N. Changes in soil chemical properties and microbial activities in response to the fungicide Ridomil gold plus copper. International Journal of Environmental Research and Public Health 2004;1:26-34.
49. Denholm I, Birnie LC, Kennedy PJ, Shaw KE, Perry JN, Powell W. The complementary roles of laboratory and field testing in ecotoxicological risk assessment. The 1998 Brighton Conferences: Pest and Diseases Volume 2. British Crop Protection Council, Farham, UK;1998. p583-590.
50. Dent D. Cultural and interference methods. In: Insect pest management, 2nd edition. CABI publishing, Cambridge, MA, USA;2000. p235-266.
51. Deribe E, Rosseland BO, Borgstrom R, Salbu B, Gebremariam Z, Dadebo E, Skipperud L, Eklo OM. Biomagnification of DDT and its metabolites in four fish Research 2010;117:51-58species of a tropical lake. Ecotoxicology and Environmental Safety 2013;95:10-18.

52. Dhaliwal GS, Singh R, Chhillar BS. In: Essentials of agricultural entomology. Kalyani Publishers, New Delhi, India;2006.

53. Dick RP. Soil enzyme activities as indicators of soil quality. In: Doran JW, Coleman DC, Bezdicek DF, Stewart BA. (eds.) Defining soil quality for a sustainable environment, Soil Science Society of America, Madison, WI, USA. 1994;p107-124.

54. Dunfield KE, Siciliano SD, Germida JJ. The fungicides thiram and captan affect the phenotypic characteristics of *Rhizobium leguminosarum* strain C1 as determined by FAME and Biolog analyses. Biology and Fertility of Soils 2000;31:303-309.

55. Dutta M, Sardar D, Pal R, Kole RK. Effect of chlorpyrifos on microbial biomass and activities in tropical clay loam soil. Environmental Monitoring and Assessment 2010;160:385-391.

56. Edwards CA. The environmental impact of insecticides. In: Delucchi V (ed.) Integrated pest management, Protection Integŕee Quo vadis? An International Perspective. Parasitis 86, Geneva, Switzerland;1987. p309-329.

57. Elbaz A, Clavel J, Rathouz PJ, Moisan F, Galanaud JP, Delemotte B, Alperovitch A, Tzourio C. Professional exposure to pesticides and Parkinson disease. Annals of Neurology 2009;66:494-504.

58. Evans SC, Shaw EM, Rypstra AL. Exposure to a glyphosate-based herbicide affects agrobiont predatory arthropod behavior and long-term survival. Ecotoxicology 2010;19:1249-1257.

59. Fabellar LT, Heinrichs EA. Relative toxicity of insecticides to rice planthopper and leafhoppers and their predators. Crop Protection 1986;5:254-258.

60. Favari L, Lopez E, Martinez-Tabche L, Dıaz-Pardo E. Effect of insecticides on plankton and fish of Ignacio Ramirez reservoir (Mexico): a biochemical and biomagnification study. Ecotoxicology and Environmental Safety 2002;51:177-186.

61. Fishel FM, Ferrell JA. Managing pesticide drift. Agronomy department. PI232 University of Florida, Gainesville, FL, USA;2013 (http://edis.ifas.ufl.edu/pi232) (accessed 14 October 2013).

62. Fishel FM. Pesticides effects on nontarget organisms. PI-85. Pesticide information office, Florida Cooperative Extension

Service, IFAS, University of Florida, Gainesville, FL, USA; 2011. (http://edis.ifas.ufl.edu/pi122) (accessed 14 October 2013).

63. Floch C, Chevremont AC, Joanico K, Capowiez Y, Criquet S. Indicators of pesticide contamination: Soil enzyme compared to functional diversity of bacterial communities via Biolog® Ecoplates. European Journal of Soil Biology 2011;47:256-263.

64. Fountain MT, Brown VK, Gang AC, Symondson WOC, Murray PJ. The effect of the insecticide chlorpyrifos on spider and collembolan communities. Pedobiologia 2007;51:147-158.

65. Garcia C, Hernandez T, Costa F. Potential use of dehydrogenase activity as an index of microbial activity in degraded soils. Communications in Soil Science and Plant Analysis 1997;28:123-134.

66. Garg, H, Atri C, Sandhu PS, Kaur B, Renton M, Banga SK, Singh H, Singh C, Barbetti, MJ, Banga, SS. High level of resistance to *Sclerotinia sclerotiorum* in introgression lines derived from wild crucifers and the crop *Brassica* species *B. napus* and *B. juncea*. Field Crops Research 2010;117:51-58.

67. Garg H, Sivasithamparam K, Banga SS, Barbetti, MJ. Cotyledon assay as a rapid and reliable method of screening for resistance against *Sclerotinia sclerotiorum* in *Brassica napus* genotypes. Australasian Plant Pathology 2008;37:106-111.

68. Georghiou GP, Taylor CE. Genetic and biological influences in evolution of insecticide resistance. Journal of Economic Entomology 1977;70:319-323.

69. Ghananand T, Prasad CS, Lok N. Effect of insecticides, bio-pesticides and botanicals on the population of natural enemies in brinjal ecosystem. Vegetos- An International Journal of Plant Research 2011;24:40-44.

70. Gianelli VR, Bedmar F, Luis J. Persistence and sorption of imazapyr in three Argentinean soils. Environmental Toxicology and Chemistry 2013. DOI:10.1002/etc.2400.

71. Gianfreda L, Sannino F, Violante A. Pesticide effects on the activity of free, immobilized and soil invertase. Soil Biology and Biochemistry 1995;27:1201-1208.

72. Giglio A, Giulianini PG, Zetto T, Talarico F. Effects of the pesticide dimethoate on a non-target generalist carabid, *Pterostichus melas*

italicus(Dejean, 1828) (Coleoptera:Carabidae). Italian Journal of Zoology 2011;78:471-477.

73. Gill HK, McSorley R, Branham M. Effect of organic mulches on soil surface insects and arthropods. Florida Entomologist 2011;94:226-232.
74. Gill HK, McSorley R, Goyal G, Webb SE. Mulch as a potential management strategy for lesser cornstalk borer, *Elasmopalpus lignosellus* (Insecta: Lepidoptera: Pyralidae), in bush bean (*Phaseolus vulgaris*). Florida Entomologist 2010;93:183-190.
75. Gill HK, McSorley R, Goyal G. Soil Solarization and organic mulches: tools for pest management. In: Dhawan AK, Singh B, Brar-Bhullar M, Arora R (eds.). Integrated pest management. Scientific Publishers, Jodhpur, India;2013. p301-324.
76. Gill HK, McSorley R, Treadwell DD. Comparative performance of different plastic films for soil solarization and weed suppression. HortTechnology 2009;19:769-774.
77. Gill HK, McSorley R. Effect of integrating soil solarization and organic mulching on the soil surface insect community. Florida Entomologist 2010;93:308-309.
78. Gill HK, McSorley R. Impact of different organic mulches on the soil surface arthropod community and weeds. International Journal of Pest Management 2011(a);58:33-40.
79. Gill HK, McSorley R. Effect of different inorganic/synthetic mulches on weed suppression during soil solarization. Proceeding of Florida State Horticultural Society 2011(b);124:310-313.
80. Gill RJ, Ramos-Rodriguez O, Raine NE. Combined pesticide exposure severely affects individual- and colony-level traits in bees. Nature 2012;491:105-108.
81. Gondhalekar AD, Scherer CW, Saran RK, Scharf ME. Implementation of an indoxacarb susceptibility monitoring program using field-collected German cockroach isolates from the United States. Journal of Economic Entomology 2013;106:945–953.
82. Gondhalekar AD, Song C, Scharf, ME. Development of strategies for monitoring indoxacarb and gel bait susceptibility in the German cockroach (Blattodea: Blattellidae). Pest Management Science 2011;67:262-270.

83. Goyal G, Kanta U. Influence of time and method of sowing of kharif maize on the incidence of *Chilo partellus* (Swinhoe). Journal of Insect Science 2005a;19:24-29.

84. Goyal G, Kanta U. Influence of barrier crops on the incidence of *Chilo partellus* (Swinhoe) in kharif maize. Journal of Insect Science 2005b;19:54-58.

85. Gonod LV, Martin-Laurent F, Chenu C. 2, 4-D impact on bacterial communities, and the activity and genetic potential of 2, 4-D degrading communities in soil. FEMS Microbiology Ecology 2006;58:529-537.

86. Grasman KA, Scanlon PF, Fox GA. Reproductive and physiological effects of environmental contaminants in fish-eating birds of the Great Lakes: a review of historical trends. In: Trends in levels and effects of persistent toxic substances in the Great Lakes, Springer, Netherlands;1998. p117-145.

87. Greenlee AR, Arbuckle TE, Chyou PH. Risk factors for female infertility in an agricultural region. Epidemiology 2003;14:429-436.

88. Grison R, Grezes-Besset B, Schneider M, Lucante N, Olsen L, Leguay JJ, Toppan A. Field tolerance to fungal pathogens of *Brassica napus*constitutively expressing a chimeric chitinase gene. Nature Biotechnology 1996;14:643-646.

89. Grover P, Danadevi K, Mahboob M, Rozati R, Banu BS, Rahman MF. Evaluation of genetic damage in workers employed in pesticide production utilizing the Comet assay. Mutagenesis 2003;18:201-205.

90. Guerrero I, Morales MB, Oñate JJ, Geige, F, Berendse F, Snoo GD, Tscharntke T. Response of ground-nesting farmland birds to agricultural intensification across Europe: Landscape and field level management factors. Biological Conservation 2012;152:74-80.

91. Gupta S, Dikshit AK. Biopesticides: An eco-friendly approach for pest control. Journal of Biopesticides 2010;3:186-188.

92. Handa SK, Agnihotri NP, Kulshreshtha G. Effect of pesticide on soil fertility. In: Pesticide residues; significance, management and analysis 1999;184-198.

93. Hargrave BT, Harding GC, Vass WP, Erickson PE, Fowler BR, Scott V. Organochlorine pesticides and polychlorinated biphenyls in the Arctic Ocean food web. Archives of Environmental Contamination and Toxicology 1992;22:41-54.

94. Hayden KM, Norton MC, Darcey D, Østbye T, Zandi PP, Breitner JCS, Welsh-Bohmer KA. Occupational exposure to pesticides increases the risk of incident AD The Cache County Study. Neurology 2010;74:1524-1530.

95. Heck JE, Charbotel B, Moore LE, Karami S, Zaridze DG, Matveev V, Janout V, Kollarova H, Foretova L, Bencko V, Szeszenia-Dabrowska N, Lissowska J, Mates D, Ferro G, Chow W-H, Rothman N, Stewart P, Brennan P, Boffetta P. Occupation and renal cell cancer in Central and Eastern Europe. Occupational and Environmental Medicine 2010;67:47-53.

96. Henry M, Bèguin M, Requier F, Rollin O, Odoux J-F, Aupinel P, Aptel J, Tchamitchian S, Decourtye A. A common pesticide decreases foraging success and survival in honey bees. Science 2012;336:348-350.

97. Hoppin JA, Umbach DM, London SJ, Henneberger PK, Kullman GJ, Coble J, Alavanja MCR, Freeman LEB, Sandler DP. Pesticide use and adult-onset asthma among male farmers in the agricultural health study. European Respiratory Journal 2009;34:1296-1303.

98. Hull LA, Starner R van. Effectiveness of insecticide applications timed to correspond with the development of rosy apple aphid (Homoptera: Aphididae) on apple. Journal of Economic Entomology 1983;76:594-59.

99. Hunter K. The poisoning of non-target animals. In: Best GA, Ruthven AD (eds.) Pesticides- developments, impacts, and controls. The Royals Society of Chemistry, London, UK;1995. p74-86.

100. Hussain S, Siddique T, Saleem M, Arshad M, Khalid A. Impact of pesticides on soil microbial diversity, enzymes, and biochemical reactions. Advances in Agronomy 2009;102:159-200.

101. Ibrahim L, Preuss TG, Ratte HT, Hommen U. A list of fish species that are potentially exposed to pesticides in edge-of-field water bodies in the European Union-a first step towards identifying vulnerable representatives for risk assessment. Environmental Science and Pollution Research 2013;20:2679-2687.

102. Imfeld G, Vuilleumier S. Measuring the effects of pesticides on bacterial communities in soil: a critical review. European Journal of Soil Biology 2012;49:22-30.
103. Ingram CW, Coyne MS, Williams DW. Effects of commercial diazinon and imidacloprid on microbial urease activity in soil. Journal of Environmental Quality 2005;34:1573-1580.
104. IRAC. Resistance management for sustainable agriculture and improved public health;2013. (http://www.irac-online.org/).
105. Ismail BS, Yapp KF, Omar O. Effects of metsulfuron-methyl on amylase, urease, and protease activities in two soils. Australian Journal of Soil Research 1998;36:449–456.
106. Jat RS, Meena HN, Singh AL, Surya JN, Misra JB. Weed management in groundnut (*Arachis hypogaea* L.) in India-a review. Agricultural Reviews 2011;32:155-171.
107. Johnson LA, Welch B, Whitfield SM. Interactive effects of pesticide mixtures, predators, and environmental regimes on the toxicity of two pesticides to red-eyed tree frog larvae. Environmental Toxicology 2013;32:2379-2386.
108. Julien MH, Griffiths MW. Biological Control of Weeds: a world catalogue of agents and their target weeds. Wallingford, Oxford: CAB International, 4th ed.;1998. p223.
109. Kelly DW, Poulin R, Tompkins DM, Townsend C. Synergistic effects of glyphosate formulations and parasite infection on fish malformations and survival. Journal of Applied Ecology 2010;47:498-504.
110. Kerby JL, Richards-Hrdlicka KL, Storfer A, Skelly DK. An examination of amphibian sensitivity to environmental contaminants: are amphibians poor canaries? Ecology Letters 2010;13:60-67.
111. Kevan PG. Pollinators as bioindicators of the state of the environments: species, activity and diversity. Agriculture, Ecosystems and Environment 1999;74:373-393.
112. Kfir R. Prospects for cultural control of the stalk borers, *Chilo partellus* (Swinhoe) and *Busseola fusca* (Fuller), in summer grain crops in South Africa. Journal of Entomological Society of Southern Africa 1990;53:41-47.

113. Khush GS, Mackill DJ, Sidhu GS. Breeding rice for resistance to bacterial blight. In: Bacterial Blight of Rice, Manila, Philippines, International Rice Research Institute1989. p207-217.

114. Kips RH. Environmental aspects. In: Haskell PT (ed.) Pesticide application: principles and practice. Clarendon Press, Oxford;1985. p190-201.

115. Kizilkaya R, Akça İ, Aşkın T, Yılmaz R, Olekhov V, Samofalovaf I, Mudrykh N. Effect of soil contamination with azadirachtin on dehydrogenase and catalase activity of soil. Eurasian Journal of Soil Science 2012;1:98-103.

116. Köck-Schulmeyer M, Ginebreda A, González S, Cortina JL, de Alda ML, Barceló D. Analysis of the occurrence and risk assessment of polar pesticides in the Llobregat River Basin (NE Spain). Chemosphere 2012;86:8-16.

117. Konar SK. Pesticides and aquatic environment. Indian Journal of Fisheries 2011;80-85.

118. Konradsen F, Van der HW, Cole DC, Hutchinson G, Daisley H, Singh S, Eddleston M. Reducing acute poisoning in developing countries-options for restricting the availability of pesticides. Toxicology 2003;192:249-261.

119. Kos M, van Loon JJ, Dicke M, Vet LE. Transgenic plants as vital components of integrated pest management. Trends in Biotechnology 2009;27:621-627.

120. Kumar M, Chimote V, Singh R, Mishra GP, Naik PS, Pandey Sk, Chakrabarti SK. Development of Bt transgenic potatoes for effective control of potato tuber moth by using cry1Ab gene regulated by GBSS promoter. Crop Protection 2010;29:121-127.

121. Kumar R, Kranthi S, Nitharwal M, Jat SL, Monga D. Influence of pesticides and application methods on pest and predatory arthropods associated with cotton. Phytoparasitica 2012;40:417-424.

122. Kumari M. Effects of organophosphate pesticide Abate on the ovary of the cat fish, *Heteropneustes fossilis* (Bloch.). Bangladesh Journal of Zoology 2012;40:207-212.

123. Laabs V, Wehrhan A, Pinto A, Dores E, Amelung W. Pesticide fate in tropical wetlands of Brazil: An aquatic microcosm study under semi-field conditions. Chemosphere 2007;67:975-989.

124. Larson SJ, Capel PD, Majewski M. Pesticides in surface waters: Distribution, trends, and governing factors (No. 3). CRC Press;2010.

125. Laycock I, Lenthall KM, Barratt AT, Cresswell JE. Effects of imidacloprid, a neonicotinoid pesticide, on reproduction in worker bumble bees (*Bombus terrestris*). Ecotoxicology 2012;21:1937-1945.

126. le Vieux PD, Malan AP. An overview of the vine mealybug (*Planococcus ficus*) in South African vineyards and the use of entomopathogenic nematodes as potential biocontrol agent. South African Journal of Enology and Viticulture 2013;34:108-118.

127. Lee DH, Lind PM, Jacobs DR, Salihovic S, van Bavel B, Lind L. Polychlorinated biphenyls and organochlorine pesticides in plasma predict development of type 2 diabetes in the elderly. The Prospective Investigation of the Vasculature in Uppsala Seniors (PIVUS) study. Diabetes Care 2011a;34:1778-1784.

128. Lee SJ, Mehler L, Beckman J, Diebolt-Brown B, Prado J, Lackovic M, Waltz J, Mulay P, Schwartz A, Mitchell Y, Moraga-McHaley S, Gergely R, Calvert GM. Acute pesticide illnesses associated with off-target pesticide drift from agricultural applications: 11 States, 1998-2006. Environmental Health Perspectives 2011b;119:1162.

129. Lee WJ, Colt JS, Heineman EF, McComb R, Weisenburger DD, Lijinsky W, Ward MH. Agricultural pesticide use and risk of glioma in Nebraska, United States. Occupational and Environmental Medicine 2005;62:786-792.

130. Lupwayi NZ, Harker KN, Clayton GW, O'Donovan JT, Blackshaw RE. Soil microbial response to herbicides applied to glyphosate-resistant canola. Agriculture Ecosystem and Environment 2009;129:171-176.

131. Madhuri RJ, Rangaswamy V. Influence of selected insecticides on phosphatase activity in groundnut (*Arachis hypogeae* L.) soil. Journal of Environmental Biology 2002;23:393-397.

132. Martinez-Toledo MV, Salermon V, Rodelas B, Pozo C, Gonzalez-Lopez J. Effects of the fungicide captan on some functional groups of soil microflora. Applied Soil Ecology 1998;7:245-255.

133. Matthews GA. Improved systems of pesticide application. Philosophical Transactions of the Royal Society B 1981;295:163-173.

134. Mayanglambam T, Vig K, Singh DK. Quinalphos persistence and leaching under field conditions and effects of residues on dehydrogenase and alkaline phosphomonoesterases activities in soil. Bulletin of Environmental Contamination and Toxicology 2005;75:1067-1076.

135. McClure MS. Resurgence of the scale, *Fiorinia externa* (Homoptera: Diaspididae) on hemlock following insecticide application. Environmental Entomology 1977;6:480-484.

136. Mcfadyen REC. Successes in biological control of weeds. Proceedings of the X International Symposium on Biological Control of Weeds 3. Montana State University, Bozeman, Montana, USA. Neal RS (ed.);2000. p3-14.

137. McSorley R, Gill HK. Introduction to soil solarization. ENY-062, Department of Entomology and Nematology, Florida Cooperative Extension Service, Institute of Food and Agricultural Sciences, University of Florida, Gainesville, FL, USA;2010. (http://edis.ifas.ufl.edu/in856) (accessed 14 October 2013).

138. Megharaj M, Boul HL, Thiele JH. Effects of DDT and its metabolites on soil algae and enzymatic activity. Biology and Fertility of Soils 1999;29:130-134.

139. Menendez A, Martínez A, Chiocchio V, Venedikian N, Ocampo J A, Godeas A. Influence of the insecticide dimethoate on arbuscular mycorrhizal colonization and growth in soybean plants. International Microbiology 2010;2:43-45.

140. Mesnage R, Clair E, de Vendômois JS, Seralini GE. Two cases of birth defects overlapping Stratton-Parker syndrome after multiple pesticide exposure. Occupational and Environmental Medicine 2010;67:359-359.

141. Meyers L, Bull J. Fighting change with change: adaptive variation in an uncertain world. Trends in Ecology and Evolution 2002;17:551-557.

142. Miedaner T. Breeding wheat and rye for resistance to *Fusarium* diseases. Plant Breeding 1997;116: 201-220.

143. Mineau P, Whiteside M. Pesticide acute toxicity is a better correlate of US grassland bird declines than agricultural intensification. PloS One 2013;8:e57457.

144. Mironidis GK, Kapantaidaki D, Bentila M, Morou E, Savopoulou-Soultani M, Vontas J. Resurgence of the cotton bollworm *Helicoverpa armigera* in northern Greece associated with insecticide resistance. Insect Science 2013;20:505-512.

145. Mitra A, Chatterjee C, Mandal FB. Synthetic chemical pesticides and their effects on birds. Research Journal of Environmental Toxicology 2011;5:81-96.

146. Mostafalou S, Abdollahi M. Concerns of environmental persistence of pesticides and human chronic diseases. Clinical and Experimental Pharmacology 2012;S5:e002. (http://dx.doi.org/10.4172/2161-1459.S5-e002).

147. Munoz-Leoz B, Ruiz-Romera E, Antiguedad I, Garbisu C. Tebuconazole application decreases soil microbial biomass and activity. Soil Biology and Biochemistry 2011;43:2176-2183.

148. Murage EW, Voroney PR, Kay BD, Deen B, Beyaert RP. Dynamics and turnover of soil organic matter as affected by tillage. Soil Science Society of American Journal 2007;71:1363-1370.

149. Murray DL, Taylor PL. Claim no easy victories: evaluating the pesticide industry's Global Safe Use campaign. World Development 2000;28:1735-1749.

150. NAS. Principles of plant and animal pest control, Vol. 3. Insect management and control. National Academy of Sciences, Washington, DC, USA;1969.

151. Niethammer KR, White DH, Baskett TS, Sayre MW. Presence and biomagnification of organochlorine chemical residues in oxbow lakes of northeastern Louisiana. Archives of Environmental Contamination and Toxicology 1984;13:63-74.

152. Niewiadomska A, Klama J. Pesticide side effect on the symbiotic efficiency and nitrogenase activity of Rhizobiaceae bacteria family. Polish Journal of Microbiology 2005;54:43-48.

153. Niewiadomska A. Effect of carbendazim, imazetapir and thiramon nitrogenase activity, the number of microorganisms in soil and yield of red clover (*Trifolium pretense* L). Polish Journal of Environmental Studies 2004;13:403-410.

154. Nowak A, Nowak J, Klodka D, Pryzbulewska K, Telesinski A, Szopa E. Changes in the microflora and biological activity of the soil during the degradation of isoproturon. Journal of Plant Diseases and Protection 2004;19:1003-1016.

155. Omar SA, Abdel-Sater MA. Microbial populations and enzyme activities in soil, treated with pesticides. Water, Air, and Soil Pollution 2001;127:49-63.

156. Ono A, Miyazaki R, Sota M, Ohtsubo Y, Nagata Y, Tsuda M. Isolation and characterization of naphthalene-catabolic genes and plasmids from oil contaminated soil by using two cultivation-independent approaches. Applied Microbiology and Biotechnology 2007;74:501-510.

157. Oppert B, Morgan TD, Kramer KJ. Efficacy of *Bacillus thuringiensis* Cry3Aa protoxin and protease inhibitors against coleopteran storage pests. Pest Management Science 2011;67:568-573.

158. Pal R, Tah J. Biodegradation and bioremediation of pesticides in Soil: Its objectives, classification of pesticides, factors and recent developments. World Journal of Science and Technology 2012;2:36-41.

159. Palmer WA, Heard TA, Sheppard AW. A review of Australian classical biological control of weeds programs and research activities over the past 12 years. Biological Control 2010;52:271-287.

160. Pan-Germany. Pesticide and health hazards. Facts and figures. 2012;1-16 (www.pan germany.org/download/Vergift_EN-201112-web.pdf) (accessed on 14 October 2013).

161. Paoletti MG. Using bioindicators based on biodiversity to assess landscape sustainability. Agriculture, Ecosystems and Environment 1999;74:1-18.

162. Parshad VR. Rodent control in India. Integrated Pest Management Reviews 1999;4:97-126.

163. Parsons KC, Mineau P, Renfrew RB. Effects of pesticide use in rice fields on birds. Waterbirds 2010;33:193-218.

164. Patnaik GK, Kanungo PK, Moorthy BTS, Mahana PK, Adhya TK, Rao VR. Effect of herbicides on nitrogen-fixation (C2H2 reduction) associated with rice rhizosphere. Chemosphere 1995;30:339-343.

165. Pelosi C, Barot S, Capowiez Y, Hedde M, Vandenbulcke F. Pesticides and earthworms. Agronomy for Sustainable Development 2013;1-30. DOI:10.1007/s13593-013-0151-z.
166. Pereira L, Fernandes MN, Martinez CB. Hematological and biochemical alterations in the fish *Prochilodus lineatus* caused by the herbicide clomazone. Environmental Toxicology and Pharmacology 2013;36:1-8.
167. Petrelli G, Mantovani A. Environmental risk factors and male fertility and reproduction. Contraception 2002;65:297-300.
168. Pettis JS, van Engelsdorp D, Johnson J, Dively G. Pesticide exposure in honey bees results in increased levels of the gut pathogen *Nosema*. Naturwissenschaften 2012;99:153-158.
169. Prasad SM, Zeeshan M, Kumar D. Toxicity of endosulfan on growth, photosynthesis, and nitrogenase activity in two species of Nostoc (*Nostoc muscorum* and *Nostoc calcicola*). Toxicological and Environmental Chemistry 2011;93:513-525.
170. Pray CE, Huang J, Hu R, Rozelle S. Five years of Bt cotton in China-the benefits continue. The Plant Journal 2002;31:423-430.
171. Price DRG, Gatehouse JA. RNAi-mediated crop protection against insects. Trends in Biotechnology 2008;26:393-400.
172. Reader SM, Miller TE. The introduction into bread wheat of a major gene for resistance to powdery mildew from wild emmer wheat. Euphytica 1991;53:57-60.
173. Reinecke SA, Reinecke AJ. The impact of organophosphate pesticides in orchards on earthworms in the Western Cape, South Africa. Ecotoxicity and Environmental Safety 2007;66:244-251.
174. Relyea RA, Diecks N. An unforeseen chain of events: lethal effects of pesticides on frogs at sublethal concentrations. Ecological Applications 2008;18:1728-1742.
175. Relyea RA, Hoverman JT. Interactive effects of predators and a pesticide on aquatic communities. Oikos 2008;117:1647-1658.
176. Relyea RA. New effects of Roundup on amphibians: predators reduce herbicide mortality; herbicides induce antipredator morphology. Ecological Applications 2012;22:634-647.
177. Relyea RA. Predator cues and pesticides: a double dose of danger for amphibians. Ecological Applications 2003;13:1515-1521.

178. Relyea RA. The lethal impact of Roundup on aquatic and terrestrial amhibians. Ecological Applications 2005(a);15:1118-1124.

179. Relyea RA. The impact of insecticides and herbicides on the biodiversity and productivity of aquatic communities. Ecological Applications 2005(b);15:618-627.

180. Richter AR, Fuxa JR. Timing, formulations and persistence of a nuclear polyhedrosis virus and microsporidium for control of the velvetbean caterpillar (Lepidoptera: Noctuidae) in soybeans. Journal of Economic Entomology 1984;77:1299-1306.

181. Richter ED. Acute human pesticide poisonings. Encyclopedia of Pest Management 2002;3-6.

182. Romero E, Fernández-Bayo J, Díaz JMC, Nogale, R. Enzyme activities and diuron persistence in soil amended with vermicompost derived from spent grape marc and treated with urea. Applied Soil Ecology 2010;44:198-204.

183. Rosell G, Quero C, Coll J, Guerrero A. Biorational insecticides in pest management. Journal of Pesticide Science 2008;33:103-121.

184. Rueda-Ayala V, Rasmussen J, Gerhards R. Mechanical weed control. In: Precision crop protection-the challenge and use of heterogeneity. Springer Netherlands;2010. p279-294.

185. Singh G, Chahil GS, Jyot G, Battu RS, Singh B. Degradation dynamics of emamectin benzoate on cabbage under subtropical conditions of Punjab, India. Bulletin of Environmental Contamination and Toxicology 2013. DOI 10.1007/s00128-013-1013-8

186. Sannino F, Gianfreda L. Pesticide influence on soil enzymatic activities. Chemosphere 2001;45:417-425.

187. Santovito A, Cervella P, Delpero M. Chromosomal aberrations in cultured human lymphocytes treated with the fungicide, Thiram. Drug and Chemical Toxicology 2012;35:347-351.

188. Saravi SSS, Shokrzadeh M. Role of pesticides in human life in the modern age: a review. In: Stoytcheva M (ed.) Pesticides in the modern world-risks and benefits. InTech;2011. p4-11.

189. Satyavani G, Gopi RA, Ayyappan S, Balakrishnamurthy P, Reddy PN. Toxicity effect of expired pesticides to freshwater fish, *Labeo rohita*. The Journal of Agriculture and Environment 2011;12:1-9.

190. Scharf ME, Meinke LM, Wright RJ, Chandler LD, Siegfried BD. Metabolism of carbaryl by insecticide-resistant and susceptible western corn rootworm populations (Coleoptera: Chrysomelidae). Pesticide Biochemistry and Physiology 1999;63:85-96.

191. Scholz NL, Fleishman E, Brown L, Werner I, Johnson ML, Brooks ML, Michaelmore CL, Schlenk D. A Perspective on modern pesticides, pelagic fish declines, and unknown ecological resilience in highly managed ecosystems. BioScience 2012;62:428-434.

192. Schreck E, Geret F, Gontier L, Treihou M. Neurotoxic effect and metabolic responses induced by a mixture of six pesticides on the earthworm*Aporrectodea caliginosa* noctuma. Chemosphere 2008;71:1832-1839.

193. Sebiomo A, Ogundero VW, Bankole SA. Effect of four herbicides on microbial population, soil organic matter and dehydrogenase activity. African Journal of Biotechnology 2011;10:770-778.

194. Shapiro-Ilan DI, Cottrell TE, Jackson MA, Wood BW. Control of key pecan insect pests using biorational pesticides. Journal of Economic Entomology 2013;106:257-266.

195. Sharma HC, Ortiz R. Host plant resistance to insects: an eco-friendly approach for pest management and environment conservation. Journal of Environmental Biology 2002;23:111-135.

196. Sharon E, Chet I, Spiegel Y. Trichoderma as a biological control agent. In: Biological control of plant-parasitic nematodes. Springer Netherlands;2011. p183-201.

197. Shepard M, Carner GR, Turnipseed TG. Colonization and resurgence of insects pests of soybean in response to insecticides and field isolation. Environmental Entomology 1977;6:501-506.

198. Shim YK, Mlynarek SP, van Wijngaarden, E. Parental exposure to pesticides and childhood brain cancer: US Atlantic coast childhood brain cancer study. Environmental Health Perspectives 2009;117:1002-1006.

199. Singh B, Mandal K. Environmental impact of pesticides belonging to newer chemistry. In: Dhawan AK, Singh B, Brar-Bhullar M, Arora R (eds.). Integrated pest management. Scientific Publishers, Jodhpur, India;2013. p152-190.

200. Singh DK, Kumar S. Nitrate reductase, arginine deaminase, urease and dehydrogenase activities in natural soil (ridges with forest) and in cotton soil after acetamiprid treatments. Chemosphere 2008;71:412-418.

201. Sofo A, Scopa A, Dumontet S, Mazzatura A, Pasquale V. Toxic effects of four sulphonylureas herbicides on soil microbial biomass. Journal of Environmental Science and Health, Part B 2012;47:653-659.

202. Solaimalai A, Ramesh RT, Baskar M. Pesticides and environment. In: Environmental contamination and bioreclamation; 2004. p345-382.

203. Soltani H, Agricultural H. Efficacy of spinosad against potato Colorado beetle *Leptinotarsa*. Hamedan Agricultural and Natural Resources Research, Hamedan, Iran;2011.

204. Son HK, Kim SA, Kang JH, Chang YS, Park SK, Lee SK, Jacobs Jr. DR, Lee DH. Strong associations between low-dose organochlorine pesticides and type 2 diabetes in Korea. Environment International 2010;36:410-414.

205. Song C, Kanthasamy A, Anantharam V, Sun F, Kanthasamy AG. Environmental neurotoxic pesticide increases histone acetylation to promote apoptosis in dopaminergic neuronal cells: relevance to epigenetic mechanisms of neurodegeneration. Molecular Pharmacology 2010;77:621–632.

206. Souza RC, Cantão ME, Vasconcelos ATR, Nogueira MA, Hungria M. Soil metagenomics reveals differences under conventional and no-tillage with crop rotation or succession. Applied Soil Ecology 2013;72:49-61.

207. Srinivasulu M, Mohiddin GJ, Madakka M, Rangaswamy V. Effect of pesticides on the population of *Azospirillum* sp. and on ammonification rate in two soils planted to groundnut (*Arachis hypogaea* L.).Tropical Ecology 2012a;53:93-104.

208. Srinivasulu M, Mohiddin GJ, Subramanyam K, Rangaswamy V. Effect of insecticides alone and in combination with fungicides on nitrification and phosphatase activity in two groundnut (*Arachis hypogeae* L.) soils. Environmental Geochemistry and Health 2012b;34:365-374.

209. Stuart SN, Chanson JS, Cox NA, Young BE, Rodrigues ASL, Fischman DL, Waller RW. Status and trends of amphibian declines and extinctions worldwide. Science 2004;306:1783-1786.

210. Sukul P. Enzymes activities and microbial biomass in soil as influenced by metalaxyl residues. Soil Biology Biochemistry 2006;38:320-326.

211. Sultana V, Ehteshamul-Haque S, Ara J, Athar M. Effect of brown seaweeds and pesticides on root rotting fungi and root-knot nematode infecting tomato roots. Journal of Applied Botany and Food Quality 2012;83:50-53.

212. Symondson WOC, Hemingway J. Biochemical and molecular techniques. In: Dent DR, Walton MP (eds.) Methods in ecological and agricultural entomology. CAB International, Wallingford, UK;1997. p293-350.

213. Tabashnik BE, Van Rensburg JBJ, Carrière Y. Field-evolved insect resistance to Bt crops: definition, theory, and data. Journal of Economic Entomology 2009;102:2011-2025.

214. Tanner CM, Kamel F, Ross GW, Hoppin JA, Goldman SM, Korell M, Marras C, Bhudhikanok GS, Kasten M, Chade AR, Comyns K, Richards MB, Meng C, Priestley B, Fernandez HH, Cambi F, Umbach DM, Blair A, Sandler DP, Langston JW. Rotenone, paraquat, and Parkinson's disease. Environmental Health Perspectives 2011;119:866-872.

215. Tavela ADO, Araújo JV, Braga FR, Silva AR, Carvalho RO, Araujo JM, Carvalho G. Biological control of cyathostomin (Nematoda: Cyathostominae) with nematophagous fungus, *Monacrosporium thaumasium* in tropical southeastern Brazil. Veterinary Parasitology 2011;175:92-96.

216. Tien CJ, Chen CS. Assessing the toxicity of organophosphorous pesticides to indigenous algae with implication for their ecotoxicological impact to aquatic ecosystems. Journal of Environmental Science and Health, Part B 2012;47:901-912.

217. Tierney KB, Baldwin DH, Hara TJ, Ross PS, Scholz NL, Kennedy CJ. Olfactory toxicity in fishes. Aquatic Toxicology 2010;96:2-26.

218. Trajkovska S, Mbaye M, Seye MG, Aaron JJ, Chevreuil M, Blanchoud H. Toxicological study of pesticides in air and

precipitations of Paris by means of a bioluminescence method. Analytical and Bioanalytical Chemistry 2009;394:1099-1106.

219. Trevisan M, Montepiani C, Ragozza L, Bartoletti C, Loannilli E, Del Re AAM. Pesticides in rainfall and air in Italy. Environmental Pollution 1993;80:31-39.

220. Tu CM. Effects of some pesticides on enzyme activities in an organic soil. Bulletin of Environmental Contamination and Toxicology 1981;27:109-114.

221. Van Leeuwen T, Vontas J, Tsagkarakou A, Dermauw W, Tirry L. Acaricide resistance mechanisms in the two-spotted spider mite *Tetranychus urticae* and other important Acari: A review. Insect Biochemistry and Molecular Biology 2010;40:563-572.

222. Vasileiadis VP, Sattin M, Otto S, Veres A, Palinkas Z, Ban R, Pons X, Kudsk P, Weide van der R, Czembor E, Moonen AC, Kiss J. Crop protection in European maize-based cropping systems: Current practices and recommendations for innovative Integrated Pest Management. Agricultural Systems 2011;104:533-540.

223. Vermeer K, Risebrough RW, Spaans AL, Reynolds LM. Pesticide effects on fishes and birds in rice fields of Surinam, South America. Environmental Pollution 1970;7:217-236.

224. Vickerman GP. Farm scale evaluation of the long-term effects of different pesticide regimes on the arthropod fauna of winter wheat. In: Greeves MP, Grieg-Smith PW, Smith BD (eds.) Field methods for the environmental study of the effects of pesticides. BCPC Monograph No. 40 British Crop Protection Council, Farnham, UK;1988. p127-135.

225. Wang LP, Shen J, Ge LQ, Wu JC, Yang GQ, Jahn GC. Insecticide-induced increase in the protein content of male accessory glands and its effect on the fecundity of females in the brown planthopper *Nilaparvata lugens* Stål (Hemiptera: Delphacidae). Crop Protection 2010;29:1280-1285.

226. Wang MC, Gong M, Zang HB, Hua XM, Yao J, Pang YJ, Yang YH. Effect of methamidophos and urea application on microbial communities in soils as determined by microbial biomass and community level physiological profiles. Journal of Environmental Science and Health Part B 2006;41:399-413.

227. Ware GW. Effects of pesticides on nontarget organisms. Residue Reviews 1980;76:173-201.

228. Weber JB, Wilkerson GG, Reinhardt CF. Calculating pesticide sorption coefficients (K sub(d)) using selected soil properties. Chemosphere 2004;55:157-166.

229. Whitehorn PR, O'Connor S, Wackers FL, Goulson D. Neonicotinoid pesticide reduces bumble bee colony growth and queen production. Science 2012;336:351-352.

230. Williams T. Silent scourge. Audubon 1997;99:28-35.

231. Winchester PD, Huskins J, Ying J. Agrichemicals in surface water and birth defects in the United States. Acta Paediatrica 2009;98:664-669. DOI:0.1111/j.1651-2227.2008.01207.x

232. Wu JY, Smart MD, Anelli CM, Sheppard WS. Honey bees (*Apis mellifera*) reared in brood combs containing high levels of pesticide residues exhibit increased susceptibility to *Nosema* (Microsporidia). Journal of Invertebrate Pathology 2012;109:326-329.

233. Wu KM, Lu YH, Feng HQ, Jiang YY, Zhao JZ. Suppression of cotton bollworm in multiple crops in China in areas with Bt toxin–containing cotton. Science 2008;321:1676-1678.

234. Xavier R, Rekha K, Bairy KL. Health perspective of pesticide exposure and dietary management. Malaysian Journal of Nutrition 2004;10:39-51.

235. Xie H, Gao F, Tan W, Wang SG. A short-term study on the interaction of bacteria, fungi and endosulfan in soil microcosm. Science of the Total Environment 2011;412:375-379.

236. Xu X, Dailey AB, Talbott EO, Ilacqua VA, Kearney G, Asal NR. Associations of serum concentrations of organochlorine pesticides with breast cancer and prostate cancer in US adults. Environmental Health Perspectives 2010;118:60-66.

237. Xuan TD, Shinkichi T, Khanh TD, Chung IM. Biological control of weeds and plant pathogens in paddy rice by exploiting plant allelopathy: an overview. Crop Protection 2005;24:197-206.

238. Yadav BC, Veluthambi K, Subramaniam K. Host-generated double stranded RNA induces RNAi in plant-parasitic nematodes and protects the host from infection. Molecular and Biochemical Parasitology. 2006;148:219-222.

239. Yasmin S, D'Souza D. Effects of pesticides on the growth and development of earthworm: a review. Applied and Environmental Soil Sciences 2010;1-9. (Article ID 678360).

240. Zhang LW, Liu YJ, Yao J, Wang B, Huang B, Li ZZ, Sun JH. Evaluation of *Beauveria bassiana* (Hyphomycetes) isolates as potential agents for control of *Dendroctonus valens*. Insect Science 2011;18:209-216.

241. Zhou X, Scharf ME, Parimi S, Meinke LJ, Wright RJ, Chandler LD, Siegfried BD. Diagnostic assays based on esterase-mediated resistance mechanisms in western corn rootworms (Coleoptera: Chrysomelidae). Journal of Economic Entomology 2002;95:1261-1265.

Chapter 4

Efficiency of Pesticide Alternatives in Non-agricultural Areas

Damien A. Devault[1] and Hélène Pascaline[1]

[1]Université des Antilles et de la Guyane, groupe Biospheres, Campus de Schoelcher, Schoelcher Cedex, France

INTRODUCTION

Global pesticide use is increasing, and such growth is recognized as stemming from agricultural needs in response to global food stress. However, pesticides are used on other areas than agricultural fields. Even if agricultural consumption of pesticides is undoubtedly the main use, the transfer from other, more impervious surfaces is regarded as a key point in understanding the fate and the global impact of pesticides, named biocides, when used for nonagricultural purposes. In the overall environment these chemicals are combined with those applied to agricultural areas, leading to confusion and thus a probable underestimation of nonagricultural pesticides. Numerous information campaigns have targeted agricultural users. The high remaining level of background contamination of rivers highlights that minoring even obliterating urban consumers precisely stultify the considered information campaigns. The ambiguous situation of port contamination will also be discussed in the present chapter.

However, nonagricultural uses mainly involve the same chemicals (e.g., herbicides) as agricultural uses. In the present chapter, the main biocides used will be listed, and then the differences in consumption depending on countries and legislation. The environmental traces of the main pesticides will be summarized with the confounding uses for watershed scale interpretation. The consequences of pesticide use depend on transfer rates, themselves conditioned by the type of surface where these chemicals are applied and their imperviousness. For highly artificialized urban areas, where biocides are mainly used, such information is pivotal: they explain a minor but significant part of aquatic environment contamination.

Alternatives to pesticide uses have been developed for decades, some even before the advent of pesticides, primarily herbicides. The present chapter will detail the alternative types, the respective efficiency depending on substratum and vegetation type. The discussion on the shortcomings of each alternative, the development level and the risks for humans or resulting from hazardous techniques (for both the environment and substratum) will distinguish promising techniques from those that have shown to be inapplicable. The authors will explain why technological impasses are patent and the possible ways to improve such technology to make them applicable. The mechanical techniques studied are mowing, brushing, rotative clogs, sweeping, and harrowing. Thermal techniques include solarization, high-pressure steam, foam mix, gas flames, infrared and to a larger extent, laser, electrocution and UV, microwave, and γ-radiation.

A SEMANTIC OBSTACLE: HOW SHOULD PESTICIDES AND BIOCIDES BE DISTINGUISHED?

Far from being trivial, this question needs to be raised prior to examining international data. Indeed, if pesticides are used for crop protection, biocides are pesticide chemicals, i.e., poisonous compounds targeting pests, but non-agricultural pests. Incidentally, molecules involved and prophylactic control molecules could be the same.

Concerning the urban context, where the excessive use of pesticides is complicated by population density and the use of pesticides by local authorities is the most aggressive which targets healthy and ostentation goals both. Pesticide inputs could combine the two environmental contamination pathways. Considering urban use except for prophylactic campaigns, pesticides are spread in kitchen gardens and around ornamentals, i.e., gardens, golf courses, parks, including plant protection. However, use of the same chemical on roads and railways, facade building protection against termites, domestic and veterinary pest control all use biocides. In the present chapter "pesticide" will be used indifferently for pesticides and biocides, unless otherwise specified.

Use of sodium chlorate and iron sulphate should be precisely evaluated and taken into account: their amount is fourfold greater than non-mineral pesticides and could explain the discrepancy of the quantities reported. As for aquatic environment contamination, this controversy seems pointless: whatever the source, the chemical impact is the sole pragmatic yardstick.

PESTICIDE USE

The Pesticides Used

First of all, inquiries on sales and applications of biocides should be considered carefully: some herbicides are indicated as being found in the urban watershed but could result from non-agricultural pesticide use in urban areas. For example, Gerecke *et al.* (2002), Devault *et al.* (2007), Gilliom (2007), and Botta *et al.* (2012) mention atrazine as consistently polluting the urban watershed without any homologation as a biocide.

Whatever the misuses, *glyphosate* and *diuron* are the most widely applied pesticides worldwide for biocide use. The main use of biocides is for weed control in developed countries, so glyphosate, whatever the geographical region, accounts for about half of non-agricultural use in terms of quantity, one-eighth of glyphosate sales (Hanke *et al.*, 2010; Blanchoud *et al.* 2007). Glyphosate is also used for agricultural ways to sign urban input, despite its increasing urban use. This trend is due to the ban of biocide use such as diuron (Okamura *et al.*, 2003; Gilliom, 2007), despite several marginal agricultural uses (vineyards and sugarcane in Australia (Haynes *et al.*, 2000)). Although diuron is banned for outdoor applications, it is found in veterinary devices and in antifouling paint additives (Irgarol 1051®), an emerging concern because of first-order kinetic façade leaching process (Burkhardt *et al.*, 2011; Wittmer *et al.*, 2011b). Even if Coutu *et al.* (2012) proposed a model integrating rainfall conditions depending on façade exposure, other climatic events (the effects of frost or sun) on building coatings and additives are not currently studied when examining the fate of façade chemicals. Non-agricultural use is reported to be equal or dominant for diuron (Bucheli *et al.*, 1998; Gerecke *et al.*, 2002; Wittmer *et al.*, 2011).

Aminotriazole (also shortened to *amitrole*) for Europe (Blanchoud *et al.* 2004 and 2007) and *prometon*for the United States (Kimbrough & Litke, 1996; Bruce & McMahon, 1996; Hoffman *et al.*, 2000; Philips & Bode, 2002, 2004, Ryberg *et al.*, 2010) are respectively the third most referenced chemical in urban areas and marginal or even ignored in the US. For Ryberg et al. (2010), prometon use may be the most widespread biocide in the US Northeast and Midwest.

In the US, urban pesticide use seems to involve other pesticides than in Europe. Braman *et al.* (1997) noted the substantial use of pendimethalin (41%) in their study area near Atlanta but no diuron or amitrole, as Glozier *et al.* (2012) also noted. Amitrole and pendimethalin are not mentioned as urban pesticides contaminating US streams by Gilliom (2007). Even if pendimethalin is known in the European market, residues of pendimethalin are not indicated in European monitoring. Similarly, prometon is indicated (Kimbrough & Litke, 1996; Bruce & McMahon, 1996; Hoffman *et al.*, 2000; Philips & Bode, 2002, 2004; Ryberg *et al.*, 2010) as a major herbicide used in the urban environment. Philips & Bode (2004) highlighted that prometon concentrations in rivers were proportionate to the population density in the corresponding watershed, Gilliom (2007) listed it as the most frequently detected of the seven herbicides used in urban areas, and Ryberg *et al.* (2010) reported that, although prometon contamination was dominant in the US Northeast and Midwest, it was the most homogeneously represented herbicide for urban areas in the country as a whole. It is the most commonly used soil sterilant in urban areas (Kimbrough & Litke); locally unavailable to homeowners, it continues to be used in these areas by licensed applicators (*Ifid.*).

Only Gerecke *et al.* (2002) mentioned DEET (insect repellent) and diazinon (used by individual gardeners). Gilliom (2007), summarizing pesticide data in US streams and groundwater, showing that, between the six most relevant insecticides, five are significantly more frequently detected in urban streams than in agricultural streams. Four of them are significantly more detected in urban areas than in agricultural areas: diazinon, carbaryl, chlorpyrifos and malathion, despite substantial climate, biocoenosis and legacy diversities. Kimbrough & Litke (1996) already highlighted that urban insecticide use was greater and less diversified than agricultural insecticide use. Consequently, urban streams are more contaminated by such chemicals than agricultural streams. Diazinon, carbaryl, chlorpyrifos and malathion were the only insecticides used in urban areas and were noted by Whitmoree*t al.* (1992) in the top 10% of the most frequently used pesticides by homeowners and certified applicators out of 312 compounds identified.

In the Croton Lake watershed, near New York City, Philips & Bode (2002 and 2004) also inventoried diazinon, in addition to carbaryl and imidacloprid. However, diazinon (with prometon) is the only one indicated as being present in densely populated watersheds (Philips & Bode, 2002, 2004). In a Californian urban context, Walters *et al.* noted carbaryl contamination due to *Homalodisca coagulata*(Say) infestation.

For urban use, Moran (2010) and Jiang & Gan (2012) reported that, in California, pyrethroids are the most widely used pesticides for urban areas. Weston & Lydy (2012) focused on pyrethroids due to their representativeness.

Considering "urban streams" as coming from watersheds whose land use was at least 25% urban and at the most 25% agricultural, Ryberg *et al.* (2010) listed pesticide residues and trends: if prometon was the herbicide the most frequently found in US rivers, herbicide trends are described as mixed. Although s-triazines are the main monitored pesticides, simazine and atrazine are more often found in rural areas. Neither a downward nor an upward trend seems to dominate, even if atrazine metabolite DEA is increasing compared to active chemicals.

Quantities

Wittmer *et al.* (2011a) noted that urban biocide consumption is within the same range as agricultural pesticides in Switzerland (1300–2000 t each), like Pissard *et al.* (2005) in Belgium. Similarly, Lassen*et al.* (2001) conclude that Denmark has high biocide consumption. This question is pivotal and could explain the clear differences between authors: Blanchoud *et al.* (2004) consider nonagricultural chemicals as approximately 1% of the total amount in the Marne River watershed (France), in accordance with several authors (Chauvel, pers. comm.), whereas Aspelin (1998) estimates it at about 25% for the US and previous authors (Lassen *et al.*, 2001; BLW, 2007; FriedliPartner *et al.* 2007) at about 50%.

Approximately 10% of pesticide quantities spread stem from nonagricultural use in developed countries (Hanke *et al.*, 2010; Kristoffersen *et al.*, 2008). Municipalities maintain recreational gardens and playing fields. Even if athletes and the young are more exposed in such places (Harris & Solomon, 1992; Bernard *et al.*, 2001; Chaigneau, 2004), this contamination pathway is not identified as a major one. Sports fields are roughly counted because of their heterogeneity: villages could present turf areas as a sports field that cannot be compared with large cities' equipment. That said, about 30,000 sports grounds have been inventoried in France: about one per town, as in all developed countries. Amenity use accounted for approximately 0.19% of pesticide use in Denmark, about 2.7% for the Netherlands and the United Kingdom, less than 3.4% for Germany, 0.6% for Finland and 1% in France (Blanchoud *et al.*, 2004).

These results, considering minor surfaces with regard to local and even global land use, is due to greater urban use of pesticides, in comparison to the same surface treated, than in rural areas (Barbash & Resek, 1996; Devault, 2007). Barbash & Resek (1998) considered that lawns received 7.4 kg/ha (insecticide: 2.4 kg/ha; herbicide: 5 kg/ha), golf courses 18.8 kg/ha (insecticide: 13 kg/ha; herbicide: 5.8 kg/ha), whereas agricultural areas received 2.3 kg/ha (0.9 kg/ha herbicide and 1.4 kg/ha insecticide).

It is also valuable to compare these results, from survey questionnaires, completed on a volunteer basis with the estimation from Aspelin (1997) and UIPP (2000): even if pesticide use is tending to decrease, other biases should be put forward: (1) hidden pesticide use such as flea collars), (2) the spontaneous trend to minimize one's own pesticide use, and (3) the lack of pesticide traceability.

Trends in Developing Countries

Developing countries' urban areas form a related context (Ecobichon, 2001), the subject of increasing concern. To provide the least expensive off-season fresh fruit (Forget *et al.*, 1993), more acutely toxic and persistent pesticides are used in developing countries (Schaefers, 1996). The trend is similar for biocide use of pesticides: pyrethroid esters are used for household spraying to repel or kill tropical disease vectors (mainly biting insects), which are nine times more expensive than DDT (Webster, 2000): without international sponsoring, poorer nations often limit or abandon control programs. Older but restricted pesticides are not patented: local or regional chemical synthesis could occur because international bans are not applied, despite the Stockholm and Basel conventions. Thus, the main pesticide intoxications occur in developing countries: data are biased by unreported cases, but the World Health Organization reported 3 million severe poisonings, including suicides and 220,000 deaths for 1990. Such results, which have since been corroborated, are caused by careless handling and home storage (under beds, on kitchen shelves (Ecobichon, 2001), lack of protective equipment (possibly due to discomfort), and individual, collective or governmental actions (Gomes *et al.*, 1999), consumption of food or beverages stored in pesticide containers for improper uses (water or food storage). Commuters may produce food in kitchen gardens but male handwork is mainly employed in cash-paying jobs in plantations surrounding cities or in industries: once planting has been completed, crop care is in the hands of women and older children, along with child care and domestic tasks. These tasks induce frequent comings and goings between indoors and the garden, enhancing pesticide exposure risks. Kitchen garden care and maintenance is so devoted to inattentive and overbusied female or infantile handwork As in developed countries, but more acutely, the long-term solution to pesticide problems is education (Ecobichon, 2000), but developing countries lack the regulatory framework, due to insufficient awareness, means and trained personnel for these controls (Ecobichon, 2001).

In all countries, more than 50% of private gardens are treated with pesticides; Hanke *et al.* (2010); in Switzerland the percentage is estimated at 90%, 60% are total herbicides for terraces and about 30% are selective herbicides for grass, shrubs and trees. Fungicides, insecticides and other pesticides (against rodents, mollusks, etc.) account for about 4% each. Thus, the main individual consumption is for esthetics, not for kitchen gardens. Twenty percent of the Swiss population spread pesticides on walkways and garden paths, although this is strictly forbidden in Switzerland (Hanke *et al.*, 2010).

In France, where about 1,100,000 ha are grassed, 605,000 ha are residential, including 23,000 ha of apartment buildings: gardens remain a status symbol. Consequently, the main grassed surface is under private control without adequate training, subjected to unclear application protocols, and receive about 5000 t of pesticides every year. For example, park treatment information is conventionally provided for a 600-m^2 applications, due to large rural gardens and parks: the indicated quantity to use could be scrupulously determined but is often interpreted incorrectly: many users only spread pesticides on a limited surface, i.e., a few square meters, but use the dose for 600 m^2 because they do not understand the instructions for use. This information base could also be lacking because 20% of individual gardeners say they are unaware of the impact of pesticides on health and the environment (French Ministry of Ecology, 2011).

In the US, Voss *et al.* (1999) identified diazinon, 2-4D, and mecoprop as the main pesticides polluting streams during rainstorms and successfully compared them to sales for residential use.

Ports and Economic Activities

Historically, human settlements were inferred to abundant and potable water resources in order to palliate technological paucity. Handworks labor and population should be supplied, resort to highly polluting techniques involved (tannery, slaughterhouses, clothiers, etc.) and wastewater treatment had not yet been invented (Leguay, 1999).

Developing landlocked cities were consequently located near large rivers, but this indispensable water could represent a major threat: even the early civilizations soon learned to protect themselves from floods. Upstream dams and channelling were beyond their ability for large streams but were rapidly set up for minor rivers.

Moreover, hydrologic droughts, historically mainly due to lack of precipitation, had dramatic consequences: even if the water supply was the main problem, maintaining a navigable depth was progressively more difficult when the size of boats increased: particularly during the 19th century, large cities accommodated their ports with low dams in order to allow barge circulation and dug artificial coves for barge mooring. Combined with the industrial era's perception of shoreline development (i.e., clear-cut logging of riparian trees), numerous cities interconnected them to an anthropized fluvial network whose shoreline erosion accounted for about half of the sediment load of urban streams (Trimble, 1997), which accumulated upstream of the urban dam.

For coastal cities, sedimentation could be due to urban activities and, as for dams, to lentic areas bought for naval security reasons. Old-named "heavens", such places could be connected to estuaries but were more often built on the shore for long-term mooring and in order to provide a calm harbor. Such conditions enhanced suspended matter deposition. Moreover, sediment could receive water or wastewater from the shore. However, boating and other naval activities induce additional pesticide consumption: antifouling is mainly performed by using pesticides against algae and shellfish. Numerous publications provide information on past and modern pesticide use, from tributyltin and its derivate to current mixes. For example, Okamura *et al.* (2003) mentioned Irgarol use in Japanese ports, where the highest Irgarol concentrations are observed. Carbery *et al.* (2006) noted the same pattern in the Caribbean harbors of the Virgin Islands and highlighted amateur mixes made with Irgarol and diuron. Their sampling included sediment, where the maximal concentrations were obviously found.

Whether river port or sea port, sediment accumulation has been observed, and sediment is very well known for accumulating metallic (Cooper & Harris, 1974) and organic (Karickhoff *et al.*, 1979) contaminants: sedimentation due to human activities induces contaminant storage in populated areas (Devault *et al.*, 2007), where pesticides are only one of several contaminants. Because of the urban context, such sediment could reach high biocide concentrations leading to contamination hotspots and, for river ports, contaminating the aquatic environment downstream during major floods.

Chauvel (2006) asked industrialists, including the transport junction, about their pesticide consumption. In descending order, industrialists consider pest control to be useful to:

- limit fire risks (herbicides against brambles and thickets), completed by the third item in this list.
- close behind fire risk, esthetic considerations are brought up: weeds are a sign of disorder, decrepitude, inactivity and, finally, abandonment. On the contrary, business and work areas have to impress competitors, customers, suppliers and employees with an image of organization, hygiene, and activity.
- Weed development could be an obstacle for rescue operations. A practical argument could be based on risks from animals on legacy obligation or on inner safety committee requirement.
- Equipment and structure alteration. Depending on the equipment and structures involved, esthetic concerns could predominate. The risk from animals is the main risk: electrical installations (power plants, airports, etc.) are sensitive to damage by animals.
- Risk of pest invasion. Some of the industrialists surveyed were in the food processing industry, but this could be redundant with the previous item.
- Health risk. Only 6% consider this risk as sufficiently pertinent to justify pesticide use (Chauvel, 2006).

ASPERSION OF PESTICIDES AND THE CONSEQUENCES OF PESTICIDE TRANSFER

In the urban context, use of aspersion depends on the substratum. Agricultural practices could be adapted to lawns and parks. To a lesser extent, clay sidewalks and paths could be treated with the same equipment. However, considering impervious substratum such as asphalt, pavement, concrete slabs or roofs (Van de Voorde, 2012), using the same techniques is not viable: urban pesticide spraying occurs in "tiger stripes" on impervious surfaces, which does not facilitate comparison with agricultural uses.

The example of railways should be cited: high-speed trains, whatever their model (the TGV in France, the Shinkansen in Japan, the ICE3 in Germany) could be struck by weeds growing on embankments and because of the high speeds attained by these trains, this could damage the rolling stock. Consequently, railway companies are identified as potentially significant polluters.

To avoid aquaplaning, rainwater should be rapidly evacuated. Roads are therefore directly connected to sewers. Even if safety imperatives prevail, this direct surface runoff could generate serious consequences (see below).

In France, 190 airports, covering between 50 and 2000 ha (Chauvel, 2006) have paved ground totaling more than 50% of nonagricultural use. Approximately 1 million km of highways and freeways cover France, combined with all types of roads covering approximately 713,500 ha, including 145,000 ha of grassed surfaces (Chauvel, 2006), about 6% of France's total surface area.

Considering railways, information is still heterogeneous except for systematic control embankments:Schweinsberg *et al.* (1999) estimated pesticide input at approximately 8–10 t/ha, but the French railway company only declares 3 kg/ha (Blanchoud *et al.*, 2004). This result highlights how linear to surface expression could bias reasoning: in France, cumulated railways are about 85,000 km long (Chauvel, 2006).

Indeed, considering maintenance of impervious surfaces, users try to control weeds growing in fissures or interfaces between impervious surfaces. This type of application also depends on fissure/interface location: along a wall, weeds could be considered as less anaesthetic or impeding than along a gutter (Zadjian *et al.*, 2004); grassed suburban sidewalks are regarded with more tolerance than city center sidewalks. Narrow cracks in the substratum are sprayed, targeting weeds, including the impervious surroundings, a practice that is more widespread than in the agricultural context, considering weed biomass as well as surfaces: a survey of the Californian Department of Pesticide Regulation (Fossen, 2009) noted that 60% of pesticide use in urban areas occurred on impervious surfaces.

Runoff Transfer

Blanchoud *et al.* (2004) estimated pesticide runoff from agricultural areas to be between 0.1% and 2.4% depending on runoff conditions. Concomitantly, under the same rain conditions, runoff in urban areas was between 0.8 and 6.7%. These results are confirmed by Wittmer *et al.* (2011a) who observed that rural pesticide runoff was between 0.4% and 0.9% when pesticide runoff in urban contexts was about 0.6–15%. The transfer rate in agricultural contexts is in agreement with Clark & Gloosby's review (2000), who estimated agricultural exportation between 1 and 4%, Leonard (1990), who estimated agricultural runoff at about 2%, and Bro-Rassmussen (1996), who determined maximum runoff in field conditions from about 0.5 to 5%. It also integrates pesticide losses from plots where storm events occurred such as highlighted by Louchart *et al.* (2001) and Revitt *et al.*, 2002. Thus, Wittmer *et al.* (2011a) propose that pesticide runoff from urban areas could be considered as one order of magnitude greater than in agricultural areas. This estimation seems to be in accordance with the literature. The agricultural maximum transfer rate observed in blind conditions (for diuron) is close to the urban maximum transfer rate, to our knowledge, at the watershed scale (45.1%, Revitt *et al.*, 2002) but is clearly much rarer than in the urban context. The 6% transfer runoff integrating agricultural and urban areas of a whole watershed proposed by Blanchoud (2001) seems to be consistent.

Apart from runoff, pesticides spread on limited-adsorption surfaces will be exposed to other processes. However, to our knowledge, no study has specifically detailed the abiotic fate of pesticides in these conditions. Indoor conditions will be detailed in another chapter.

Lixiviation Transfer

Less studied and less obvious, the impact of urban areas on lixiviation remains significant (1) because the lixiviation volume is minimized and (2) because pesticide transfer to groundwater differs comparatively to other land uses (Trauth & Xanthopoulos, 1997).

As previously indicated, water cycles in urban areas are modified: the contribution to groundwater is halved compared to the natural water cycle. Compared to urban pesticide use, groundwater could be more contaminated than under agricultural land. Thus, it is possible to identify the urban impact on groundwater just as it is possible to identify the urban impact on surface waters.

Bruce *et al.* (1996) distinguished residential, commercial, and industrial areas in the urban impact on groundwater. Commercial areas have a greater impact on groundwater because of ornamental plants as well as roads and parking lots, while residential areas are more marked by the needs of ornamental plants and industrial areas by impervious surfaces. Residential areas showed higher contamination levels than industrial areas. However, Trauth & Xanthopoulos examined this segregation: urban areas mix industrial plants (i.e., point source contamination), roads and sewers (linear contamination), and allotment areas (surface contamination): groundwater contamination is not the faithful reflect of the surface one. Nevertheless, statistical results on studies on wells have shown that pesticide concentrations were higher in urban areas than in rural areas. Even if some pesticides are found more frequently in urban areas, statistical consistence is impacted by the number of wells. Malaguerra *et al.* (2012) outline groundwater contamination via the groundwater table and sediment from contaminated streams caused by enhanced runoff in urban areas. Inversely, because of less vertical water transfer due to impervious surfaces, leaching could be slowed, favoring degradation and lateral water transfer, mixing groundwater contaminants (Malaguerra *et al.*, 2012).

RESIDENT EXPOSURE TO PESTICIDES SPREAD IN URBAN AREAS

Pesticide use in urban areas is a major concern for the aquatic environment as well as for human health (Van Maele-Fabry *et al.*, 2011). The influence of water contamination resulting from urban pesticide runoff is greater on an aquatic environment than on human health, and food contamination is due to agricultural applications of pesticides. Consequently, the main exposure of urban residents by pesticide spread in urban areas stems from air contamination (Ragas *et al.*, 2011). Moreover, except for pesticide use in urban areas compared to agricultural areas, the urban context favors human contamination by atmospheric pesticides. Due to hydrophobic patterns of the majority of pesticides, contamination by dust is the main source of contamination by air. The aim of the present chapter is not to propose a review of the abundant literature on contamination by pesticides and associated dusts. Appropriate reviews exist, e.g., Schneider *et al.* (2003), Bradman & Whyatt (2005), Garcia-Jares (2009) Kanazawa & Kishi (2009) and Karr (2012). The aim is rather to explain why the urban context encourages human contamination.

In short, buildings are enclosed, windless, sunless, partly septic spaces where dusts can be trapped and accumulate, particularly in fabrics such as carpets (Obendorf *et al.*, 2006): 80% of pesticides found indoors are found in clothes, particularly shoes (Quiros-Alcala *et al.*, 2011). Moreover, degradation occurs less indoors than outdoors (Roberts *et al.*, 2009). Other variables independently associated with dust levels included temperature and rainfall, storing pesticide products in the house, housing density, imperfect housecleaning, and air conditioning (Harnly *et al.*, 2009). Farmworkers expose their families more than other professional categories (Quiros-Alacala *et al.*, 2011); consequently, in suburbs, municipal service employees and private gardeners could be considered as possible vectors to their relatives. Weschler & Nazaroff (2008) outlined the relationships between gaseous organic chemicals, including pesticides, and dust contamination: solubility and Koa (partition coefficient between octanol and air for chemicals) successfully describe gaseous pesticide contamination, in contrast to other molecules (Schoeib *et al.*, 2005). Clothes abrasion and other contaminations (paint coating) occur indoors, promoting indoor pesticide content and some of the organic matter in dust, such as cotton linters, may differ substantially from octanol in terms of sorption of gas-phase Semi-Volatil Organic Compounds (Weschler & Nazaroff, 2008). Direct contact of dust with polluted surfaces seems to be enough to pollute dust (Clausen *et al.*, 2004). Moreover, high indoor temperatures induce chemical volatilization, and the difference between the laboratory temperature for Koa determination and the private indoor temperature could be significant. Passive air sampling does not efficiently inform about long-term contamination because of passive samplers (Weschler & Nazaroff, 2008) and quantification thresholds.

Blanchoud (2001) estimated agricultural pesticide amounts used on the Marne watershed at about 5200 t/year, urban use at about 62.5 t/year and atmospheric amounts at about 0.5 t/year. But global contamination should not be ignored: MCE (2003) estimated that Rhine valley inhabitants, by breathing, were twice as contaminated by pesticides than if they drank 1.5 L of water with close to 0.5 μg/L total pesticide concentration, i.e., the maximum allowed concentration by surface water quality norms. Moreover, gaseous pesticides are directly bioavailable compared to pesticides associated with particles, which sequester more than 99% of the main pesticides (Koc and Kow>2). Studies on pesticide exposure mainly target farmers and pesticides used in agricultural areas (Mercadante *et al.*, 2012). Considering the issues at hand, data on public exposure to urban pesticide use are rare, even if studies are currently in progress.

Population exposure to contaminated particles or volatile pesticides is more than ever an issue because this exposure occurs as much at home as at work, and because enclosed living spaces affect every age group.

Air contamination data is still too rare and incomplete, and would benefit from further study.

To pesticides designed to protect crops, one must add a large number of biocides designed for heath or esthetic uses (household products, paints containing algaecides, etc.): the nature of these pesticides is not well understood by users. Because the sense of sight prevails over the other senses, the most readily perceived pollution is air pollution, associated with transport. Coupe *et al.* (2000) assert that high oxidative conditions in urban areas compared to rural areas (Finlayson-Pitts & Pitts, 1986) promote pesticide oxidative degradation. There is no evidence of a significant urban influence.

PROGRESSIVE PESTICIDE AWARENESS OF URBAN PESTICIDE USE

For Denmark and the Netherlands, the first monitoring programs demonstrated evidence of water contamination. Depending on its groundwater for drinking water, in 1995 Denmark discovered its groundwater pollution level. For the Netherlands, water pollution was striking because of the Meuse River contamination, which resulted in a ban, forbidding water intake for 7 weeks (1993 and 1994), while this country depends on surface water for 40% of its drinking water, soon to rise to 50%. In wooded Sweden, the threat to human health was the main driver because Swedish forestry and roadway services air-applied Agent Orange, a 2,4-D and 2,4,5-T formulation known for its mutagenic potential. Concerned by Agent Orange use and air-spraying, the population continued to debate about the daily place of pesticides after Agent Orange's definitive ban (1977). The first monitoring campaigns were carried out in 1985 and revealed water resource contamination, leading to early and radical directives (Ulén *et al.*, 2002).

Pesticide awareness differed in the largest countries. For agricultural countries such as the United States and France, pesticide awareness came early but was mainly associated with agricultural use. In Germany, the negative effects of pesticides were avoided by early plant protection legislation: the first legislative provision was decreed in Germany in 1919 and was implemented in 1968. Thus, weed control by herbicides is forbidden on hard surfaces without local authorization and only if there is no runoff risk. In this case, plant protection control programs determine the few available chemicals. For the US and France, pesticide contamination evidence dates back to the 1960s. Associated with agriculture, pesticide use was rarely reported in urban areas until the 1990s when extensive monitoring, directives in other countries, and early scientific publications (Cole *et al.*, 1984) awakened awareness. The Nationwide Urban Runoff Program, prepared between 1978 and 1980, carried out between 1980 and 1982, provided the first public information on urban contamination in the US. In comparison, the first French publication on urban pesticide contamination is Chevreuil *et al.*

(1996), and was still focused on agricultural contamination; even if the Water Law was decreed in 1992, compatible urbanization was taken into account in 2004 (Diren, 2010). The European Union is a key factor in French pesticide awareness. Like the above-mentioned countries, the UK did not experience a catalyst event leading to massive pesticide awareness. Without previous legislation as in Germany, and without agriculture importance like in the US and France, urban pesticide use would have been more evident given Greater London's importance in UK land use. However, British awareness seems limited and related publications are scarce (only Rule *et al.*, 2006, and Stuart *et al.*, 2012).

The virtuous pesticide approach is performed in agreement with the European Union: at the same time, legislators follow European directives such as the Council Directive 91/414/EEC (EEC 1991) and the Water Framework Directive (EEC, 2000), and support forums on amenities or pesticide representations. However, this process is more efficient in countries with leading governance such as France: British self-regulation practices and the tradition of voluntarism make them less easy to apply (Grundy, 2007).

Initiatives could be combined as is done in the US. First of all, professionals are involved, then private applicators. Consequently, since 1993, the US has established a 2-year license for spraying pesticides depending on member states' initiatives, presenting different process levels. For example, Idaho, Georgia, and Minnesota have established a voluntary program for the publication of a pesticide sale and use database. Idaho and Georgia follow Urban Pest Management Programs in order to involve individual applicators.

Although Germany's 1919 plant protection decree was a notable base for environment protection, the country continues to strengthen its pesticide reduction policy even if detailed data for urban herbicide do not exist. Parenthetically, annual consumption of pesticides used in part in urban areas is about 230 t. Kristoffersen *et al.* (2008) estimated glyphosate, the main pesticide used in urban areas, at about 2 t/year. Finland's use is estimated at about 5–6 metric tons of active molecules per year (*Ibid.*); the substances allowed are limited and use of very toxic pesticides is limited to qualified persons.

Pesticides used in urban areas are limited; for example, diuron is often forbidden in Europe Union countries and the Canadian province of Ontario (decree 63/09, 4 March 2009, enforced 22 April 2009) banned all pesticides use for esthetic purposes, but some limited uses, such as on golf courses, are allowed. Golf courses require intense pest control and artificialized surfaces, with pesticide transfer close to urban areas. In France, greens cover about 20,000 ha. Considering the 550 golf courses in France, the average surface of a green is about 36.4 ha. Even if economic arguments, ecological concerns, and society's growing awareness are influencing golf course managers taking these concerns into account, the results from such sites should be considered with caution, due to divergent goals or the risk of different interpretations.

The ultimate level of urban pesticide use awareness is differential taxation and alternative innovation. The Netherlands and Denmark are the most forward-looking countries for alternative development, followed by Sweden, the leading European countries for environmental issues.

In addition to legislation, the importance green political parties or related are a better reflection of the population's awareness of ecological concerns, as expressed in elections in the number of deputies for a given population: Germany sent 22 ecologist deputies to the European Parliament, France 19, Sweden, the Netherlands and the United Kingdom 5, but, in contrast to others, the United Kingdom's deputies are mainly autonomists. With these results, the political interest in the environment could be considered as moderate (Kristoffersen *et al.*, 2008). Thus, adoption of an ethical attitude will be limited until citizen support is expected. For example, despite its legislation strictly controlling urban pesticide use since 1954, Finnish people do not show a willingness to complete its legislation by greater amenity pesticide control (Kristoffersen *et al.*, 2008). Moreover, hard surfaces or the status of amenity areas could curb initiatives: in the United Kingdom, administrative land fragmentation results in local authorities being responsible for weed control (Grundy, 2007).

TECHNOLOGIC ALTERNATIVE

Road shoulders were mowed in certain places up to the 1950s, although hay production was declining at this time. Green shoulders limit soil erosion and consequently prevent road sap, helps drivers see curves and anticipate the course of the road, allows a good visibility of signs, protect from wind, and prevent monotony for drivers and eyesores for residents. However, walkers, wildfowl and rain require road shoulders to be flush cut. Margoum (2003) highlighted how ditches could enhance pollutant retention. Considering pesticide costs and low user solicitation, highway companies could notably reduce their biocide budget using alternatives to pesticides (at least 50%, Mahe (2007)).

Based on Table 1 (Marque & Chabaud, 2006), in order to control at least 8- to 10-cm-high weeds, mowing seems to be the best alternative. Mowing does not induce soil or root lifting, and cutting at an appropriate height could avoid passage: flush cutting weeds too short could harm low-growth perennial plants, which inhibit high-growth annuals such as allergen ambrosia. Moreover, perennial weeds are often more endemic than annual weeds, contributing to biodiversity promotion. Mowing seems to correspond to private and public professionals' financial means and satisfaction surveys highlight its popularity (Mahe, 2007).

In Germany, a system has been developed, the Rotofix, a hand-operated roller sprayer as an alternative to spray a zone for a single plant. Appropriately used, it could reduce herbicide volume by 75–95% (Hermanns *et al.*, 2006).

Instead of using pesticides, public authorities could employ other molecules, such as acetic, citric and pelargonic acids on hard surfaces. Although it is used in Germany, acetic acid is also prohibited in 50% of Swedish municipalities and is only allowed in the Netherlands when there is no runoff risk.

Ground cover could be an alternative (Table 1), if the ground use is amenable (Marque & Chabaud, 2006). Mulching and covering with plastic drastically limit weed growth, but could be too aleatory (vegetal wood and cloth covers), temporary (vegetal and cloth), fire-prone (vegetal, polypropylene and cloth), unaesthetic (polypropylene, vegetal cover with time), difficult to deploy (minerals are heavy, vegetal covers need time), and require maintenance (vegetal and mineral covers).

Table 1: Qualitative and economic analysis of alternative preventive methods (from Marque and Chabeaud, 2006)

Technique	Material	Advantages	Disadvantages	Raw material cost	Productivity cost	Number of input (/year)	Cost/m²	Durability
Mulching	Vegetal matter	Ease of use Natural visual Water protection Matter organic input	Lengthy deployment; Volatile; Animal deterioration; Annual complement; Visual with time; Risk hazard; Dubious herbicide	0.2 to 2€/m²	3€/m²	0.1*	3.3 to 6.2€/m²	1 to 5 years
Mulching	Mineral matter	Heavy products	Ease of use but lengthy; Water protection; Animal deterioration; Needs complement; Herbicide	2.5 to 3€/m²		1*	5.5 to 6€/m²	5 years et +
Mulching	Plastic matter, polypropylene	Easy and fast to deploy; Water protection	Impervious; Visual controversy; Film eviction problems Risk hazard; Herbicide	0.25 to 0.70€/m²	0.70€/m² for plantation + for eviction	0.25 to 0.15*	0.95 to 1.40€/m²	5 to 7 years
Mulching	Ligneous slabs	Ease of use	Impervious; Dubious herbicide	0.5 to 2.5€ each	0.60€ to 1.10€ / each	1	1.10€ to 3.60€/m²	2 to 4 years
Mulching	Plant fibers and rags	Ease of use; Water protection; Recycling	Fragility products; Visual; Risk hazard; Dubious herbicide	1 to 4€/m²	1€/m²	0*	2 to 5€/m²	2 to 5 years
Ground cover	Plants	Natural; Esthetic; Good cover with time	Needs protection the first years; Maintenance	10m€/m²	3.5€/m²	Retouches*	13.50/m²	10+ years
Chemical	Anti-germinatives	Easy for granule; Fast use; Efficient;	Pesticide use; Follow good practices and framework legislation;	0.05€/m²		1	0.11€/m²	1 year
	Foliate	Selective products	Qualified staff	0.01€/m²	0.06€/m²	1 to 3	0.07€ to 0.21€/m²	Following regrowth

Mechanical Alternatives

Brushing can be used only on impervious and clean surfaces. At best, the rotation of the bristles extracts part of the roots. However, coated surfaces are abraded: asphalt near fissures could be snatched, deteriorating bristles. Pavements should be cohesive and regular but slipping could occur when wet. Moreover, steel brush tests have demonstrated the level of noise and vibration is incompatible with good working conditions and urban use (Hansson *et al.*, 1992 in Rask & Kristoffersen, 2007). Brushes are only made in polypropylene. Despite brushing's efficiency, Lefevreet *al.* (2001) and Wood (2004) do not recommend it for long-term use but Lefevre *et al.* (2001) andHein (1990) propose to use it for heavily weeded areas.

Rotative clogs comprise a heavy metal cylinder rolling on the ground and extracting roots. They are only used on pervious surfaces, which should be tamped after the application, an expensive step. The surface is severely abraded: rotative clogs could be limited to clay surfaces (Hamelet, 2004).

Sweeping, whether or not it is mechanized, even in gutters, could be useful, despite the number of sweepers required, and is a no n-hermal alternative (Hein, 1990; Parker & Huntington, 2002; Hansen, 2004): the advantages of cleaning could justify the price of optional engines or numerous teams.Lefevre *et al.* (2001) considered that seven to ten operations per year were very efficient for controlling weeds in temperate climates.

Harrowing is still efficient on gravel surfaces: easy to use, inexpensive to purchase, maintain, and deploy, in 1992 it led to banning herbicides for churchyard treatment in Denmark (Tveedt *et al.*, 2002, in Rask & Kristoffersen, 2007).

Paradoxically, human mechanical work, whether it is used marginally or institutionally, seems to keep up for limited surfaces (Angoujard *et al.*, 1999): the Versailles municipality organizes hoeing teams of seasonal workers (Mahe, 2007). The main obstacle is the cost of labor for developed countries with high labor costs, but this obstacle could be reduced in emerging or developing countries where sweeping appears to be a reliable alternative to herbicide use. However, such practices could be considered as retrograde and even degrading.

Thermal Alternatives

Thermal alternatives use heat to scorch or burn off weeds. Heat could be obtained with sun (solarization), high-pressure steam, sugar foam, infra-red, freezing or gas flames (Table 2).

Table 2: Qualitative and economic analysis of alternative curative methods (from Marque and Chabeaud, 2006). *Short-term rental including equipment+driver/technician **Equipment Rentals vehicle or technician without applicator ***If mechanized implementation

Technics	Material	Advantages	Disadvantages	Investment in material	Productivity	Number of passages	Annual cost	Consumable
	Hoe	Ease of use; All weather; Handy; Very low investment	Number of passage; Labor-intensive; Limited to small surfaces	20 to 30€	50m²/h	5 to 6	0.40€/m²	
	Rotative clog	Low investment; Ease of use	Number of passages; Only on pervious surfaces, even degradation					Gasoil
	Rotative brush	Moderate investment; Ease of use; Effective on pavement	Substratum and joint degradation; Number of passages	4k to 5k€	1000m²/h	4 to 6	0.36€/m²	Gasoil
Mecha-nical	Sweeper	Cleaning; Sharable cost	Average to very high investment; Number of passages; Joint degradation; Limited to gutters	2,65k to 90k€	2600m²/h	8 to 12	0.12 €/m²	Gasoil
	Mowing, small equipment	Moderate investment; Ease of use; All weather; Handy	Number of passages; Labor-intensive; Not for large surfaces; Fossil-energy dependent	200 to 600€	450 m²/h	3 to 4	0.23 €/m²	Two-stroke gasoline
	Mowing/ mulcher, big equipment	All weather; Handy; Effective on ligneous	Average to expensive investment; Consumption of fossil energy; Visual on ligneous	2 to 20k€	3 to 6000 m²/h	1 to 2	0.12 €/m²	Gasoil
	Infrared	Medium investment; Ease of use; Handy; Useful on pervious as impervious; Manual or mechanized	Number of passages; Risk hazard; Only for seedlings and monocotyledons; Consumption of fossil energy; Greenhouse gases	3.3k to 9.5k€	2 km/h 1000m²/h	6 to 8	0.24 €/m²	Propane (1kg of gas/h/burner)
	Flame	Low to moderate investment; Ease of use; Manual or mechanized	Number of passage; Risk hazard; Consumption of fossil energy; Greenhouse gas generation	0.5 to 6,2€	3 to 5 km/h	4 to 6	0.22 €/m²	propane 2kg of gas/h/burner
Thermal	Steam	Efficient on impervious surface; Handy; Mechanized; Polyvalent; Cleaning effect; Disinfection	Expensive investment and water; Moderate efficacy on settled weeds; Number of passages; Diesel fuel consumption; Ground speed 0.7 to 1 km/h	17.5k to 44k€	1600 m²/h 1100 m²/h	3 to 6	0.25 €/m²	Eau (4 to 500 L/h) gasoil; Gasoil (4 to 7L/h)
	Warm water	All weather; Handy; Polyvalent; Hose; Effective on bryophytes	Average investment; Number of passages; Only on leaves; Water and diesel fuel consumption; Ground speed 1 to 2 km/h	15k to 18k€	1000 m²/h	4 to 8	0.25€/m²	Water (3 to 400 L/h) gasoil (4 to 5L/h)
	Warm foam	Useful on pervious and impervious surfaces; Handy; All weather; Number of passages	For rent only; Mainly effective on leaves; Water consumption; Diesel fuel consumption; Ground speed, 3 to 5 km/h	0.7k€/d*; 0.9k€/w; 3k€/m; 13k€/6m; 21k€/y; 48k€/4y	350 m²/h	3 to 4	1€/m²	Water (15 to 50m³/ha); Gasoil (5 to 6L/h); Foam 100€/d; 0,2 to 0,4% additive
	Foliate	Low to moderate investment; Ease of use; Pervious and impervious surfaces; Manual or mechanized; Dose modulation following the weeds	Weather dependent; Run-off risk on impervious surfaces; Number of passages; Average efficacy on certain plants	50 to 3k€	1000 m² to 2000 m²/h	3 to 5	0.11 to 0.17 €/m²	Products 35€ Water Gasoil ***
Chemical	Foliate + anti-germinative	Low to moderate investment; Number of passages; Manual or mechanized; Curative and preventive	Weather dependent; Pervious surfaces only	50 to 3k€	1000 m² to 2000 m²/h	1 + 1 retouch	0.12 to 0.18 €/m²	Products 400€ Water Gasoil ***
	Foliate-residuary			50 to 3k€	1000 to 2000 m²/h	1 + 1 retouch	0.15€/m²	Products 150€ Water; Gasoil ***

Globally, thermal uses require many passages (Rask, 2012) and are highly energy-consuming. Treatments are more effective on low-growth weeds and roots are scarcely damaged. The driving speed must be slow for an effective treatment. The main target of thermal alternatives is to expose pesticides to warm conditions, so as to degrade them. However, especially when the vector of the fluid, i.e., steam, warm water or a warm mix, the temperature reached should not be high enough to degrade or even mineralize the pesticide, but could enhance volatilization. This phenomenon is known to significantly affect the fate of some pesticide families. In the urban context, due to a lesser adsorption phenomenon leading to enhanced pesticide runoff, ground temperature, H Henry's constant greater than 10^{-5}, and the effects of Raoult's law, pesticides could be exposed to enhanced volatilization (Burkhardt & Guth, 1981). For Scheyer *et al.* (2007ab) and Delaunay *et al.* (2010), high amounts of volatilized urban pesticides are notably observed in urban air but are too limited to induce long-distance contamination or to significantly pollute agricultural areas when farmland pesticides are found on the same order of magnitude in agricultural areas as in urban areas.

Considering *solarization*, two limits are identified. First of all, the weather should be sunny (at least 250 h/month) and shade should be avoided (due to other weeds). Also, a large amount of plastic waste is generated (Cheroux & Serail, 2006).

Due to the nature of impervious surfaces, i.e. mainly dark asphalt, solarization could lead to extreme temperatures (asphalt fusion temperature: between 90°C and 110°C). Many pesticides, particularly herbicides, could lyse at such temperatures, but no study has investigated this question. Even if the sunshine does not induce high temperatures, photolysis could occur but no direct evidence has been found in the literature for this special case. However, the long-term experiments conducted byJorgenson & Young (2010), Jiang *et al.* (2011) and Jiang & Gan (2012) do not mention photolysis of urban pesticide, but experiments examined low photolysis-sensitive pyrethroids. However, the observed loss is far from being as fast as expected with less than 1 h DT50 photolysis at neutral pH (Fossen, 2006).

High-pressure steam application requires substantial quantities of water and a substantial financial investment. Its efficacy is poor (Daar, 1994) because of perennials. Foam could be used instead of water, made of coconut sugar and corn sugar, to enhance warming duration and subsequent efficacy (Daar, 1994). Numerous applications are required.

Gas flames alternative use has the advantage of being an intuitive and light (Rask, 2012). However, this alternative is expensive (substantial gas consumption) and may even be a source of fire danger (Wood, 2004). It is the most commonly applied thermal weed control method on hard surfaces. In Germany, a train equipped with flame weeders has been elaborated for railway embankments (Kreeb & Warnke, 1994).

Infrared radiation is the most effective non-chemical control method and economically comparable to herbicide treatment (Augustin, 2003, cited by Rask & Kristoffersen, 2007), but radiators are very expensive, brittle, and inoperative for dense vegetation (Ascard, 1998).

Other alternatives have been tried: laser (from Couch & Gangstad, 1974, to Heisel *et al.*, 2002), gamma radiation (radioactive), UV radiation (nullified by mutagenic and fire hazards), microwaves (hazardous and need 1000–3400 kg diesel/ha for a significant effect according to Sartorato *et al.*, 2006), electrocution (fire hazard for the surrounding terrain, electrocution risk for operators and passers-by, and a high amount of electricity needed, but this could be an alternative for railway pesticide uses). None of these methods currently presents consistent results.

CONCLUSIONS

The use of pesticides is still too directly associated with agriculture, a clear cultural barrier for those countries that are built on a strong dichotomy between the countryside and the city, the latter needing to be maintained. However, in developed countries, urban sprawl seems to be the main driver of water resource contamination. Outer urban areas are the most vulnerable to pesticide use: they do not have a water collection and treatment system as developed as those in city centers, while suffering from a greater level of pesticide pressure than is found in agricultural areas. However, studies on pesticide representation are mainly done in agricultural areas, and their urban equivalents are rarer.

Populations believe that the impact of indiscernible actors of pollution does not exist. Moreover, the fate of pesticides in urban substrates (asphalt, etc.) is not sufficiently known: recent studies on contaminated concrete and paint have shown gaps in our understanding of sorption, volatilization, and photolysis processes.

The descriptors of the pesticide pressure are lacking but also in need of improvement: comparing the surfaces covered by agricultural machinery spraying to urban "leopard spot" or "tiger stripe" spreading is not relevant. This lack of standardization can be found in studies seeking to highlight the performance of alternative techniques for spreading, which nevertheless seem capable of improvement.

Nonetheless, new curative or preventive tools could provide effective alternatives to pesticide use. The pivotal quality of alternative strategies lies in the choice of matching the tool to the substratum. However, this pas de deux is essential for limiting pesticide contamination.

REFERENCES

1. Angoujard G., Ferron O., Blanchet P., 1999. Désherbage des surfaces non végétalisées. Evolution des techniques et des produits en espaces verts. Phytoma-La Défense des Végétaux, 517, 25-27. *In French with English summary*
2. Ascard, J., 1998. Comparison on flaming and infrared radiation techniques for thermal weed control. Weed Research 38, 69-76.
3. Aspelin, A.L. 1998. Pesticides Industry Sales And Usage: 1994 and 1995 Market Estimates. U.S. Environmental Protection Agency, Washington, DC.
4. Barbash, J.E., and Resek, E.A., 1996, Pesticides in ground water—Distribution trends, and governing factors: Ann Arbor Press, Chelsea, Michigan, p. 3
5. Bernard C., Nuygen H., Truong D., Krieger R. 2001. Environmental residues and biomonitoring estimates of human insecticide exposure from treated residential turf. Archives of Environmental Contamination and Toxicology 41, pp. 237–240.
6. Biros, C., 2011. Contribution à l›étude du discours environnemental : les organisations et leurs discours au Royaume-Uni. Ph.D Thesis. Université Bordeaux 2. Bordeaux, 643 pp. *In French.*
7. Blanchoud, H., Moreau-Guigon, E., Farrugia, F., Chevreuil, M., Mouchel, J.M. 2007. Contribution by urban and agricultural pesticide uses to water contamination at the scale of the Marne watershed. Science of the Total Environment 375 (1-3), pp. 168-179.
8. Blanchoud, H., Farrugia, F., Mouchel, J.M. 2004. Pesticide uses and transfers in urbanised catchments. Chemosphere 55 (6) , pp. 905-913.
9. Bradman, A., Whyatt, R.M., 2005. Characterizing exposures to nonpersistent pesticides during pregnancy and early childhood in the National Children's Study: A review of monitoring and measurement methodologies. Environmental Health Perspectives 113 (8) 1092-1099.
10. Bruce B.W., McMahon P.B. 1996. Shallow ground-water quality beneath a major urban center: Denver, Colorado, USA? Journal of Hydrology 186, 129-151.

11. Botta F., Fauchon N., Blanchoud H., Chevreuil M., Guery B. 2012. Phyt'Eaux Cités: Application and validation of a programme to reduce surface water contamination wuth urban pesticides. Chemosphere 86, 166-176.
12. Bucheli, T.D., Müller, S.R., Heberle, S., Schwarzenbach, R.P., 1998. Occurrence and behavior of pesticides in rainwater, roof runoff, and artificial stormwater infiltration. Environmental Science and Technology 32 (22) 3457-3464.
13. Burkhard N., Guth J.A. 1981. Chemical hydrolysis of 2-chloro-4,6-bis (alkylamino)-1,3,5-triazine herbicides and their breakdown in soil under the influence of adsorption. Pesticide Science 12, 45-52.
14. Burkhardt M., Junghans M., Zuleeg S., Boller M., Schoknecht U., Lamani X., Bester K., Vonbank R., Simmler H. 2009. Biocides in building facades - Ecotoxicological effects, leaching and environmental risk assessment for surface waters. Umweltwissenschaften und Schadstoff-Forschung 21, 1, 36-47.
15. Burkhardt M., Zuleeg S., Vonbank R., Schmid P., Hean S., Lamani X., Bester K., Boller M. 2011. Leaching of additives from construction materials to urban storm water runoff. Water Science & Technology 63, 9, 1974-1982.
16. Chaigneau A. 2004. Pesticide transfer used on grass in France and sportsmen's contamination. Crop protection service report, 66pp. *In French*
17. Chauvel, G. 2006. Les Zones non agricoles. http://www.environnement-haute-garonne.fr/annexe1_Chauvel_12052009.pdf. Acceded January 20 2013. 41pp. *In French*.
18. Chevreuil M., Garmouma M., Teil M.J., Chesterikoff A. 1996. Occurrence of organochlorines (PCBs, pesticides) and herbicides (triazines, phenylureas) in the atmosphere and in the fallout from urban and rural stations of the Paris area. Science of the Total Environment 182, 25-37.
19. Clark G.M., Goolsby D.A. 2000. Occurrence and load of selected runoff components. Science of the Total Environment 193, 215–228.
20. Phillips, P. J., G. R. Wall, E. M. Thurman, D. A. Eckhardt, J. VanHoesen. 1999. Metolachlor and its metabolites in tile

drain and stream runoff in the Canajoharie creek watershed. Environmental Science and Technology 33, 3531–3527.

21. Clausen, P.A., Hansen, V., Gunnarsen, L., Afshari, A.,Wolkoff, P., 2004. Emission of di- 2-ethylhexyl phthalate from PVC flooring into air and uptake in dust: Emission and sorption experiments in FLEC and CLIMPAQ. Environmental Science and Technology 38, 2531-2537.
22. Cole, R.H., Frederick, R.E., Healy, R.P., Rolan, R.G., 1984. Preliminary findings of the priority pollutant monitoring project of the Nationwide Urban Runoff Program. Journal of the Water Pollution Control Federation 56, 7, 898-908.
23. Cooper, B.S., Harris, R.C., 1974. Heavy metals in organic phases of river and estuarine sediment. Marine Pollution Bulletin 5 (2) 26-27.
24. Couch R, Gangstad E O. (1974). Response of waterhyacinth to laser radiation. Weed Science 22 (5), 450-453
25. Coupe R.H., Manning M.A., Foreman W.T., Goolsby D.A., Majewski M.S. 2000. Occurrence of pesticides in rain and air in urban and agricultural areas of Mississipi, April-September 1995. Science of the Total Environment 248, 227-240.
26. Coutu, S., Del Giudice, D., Rossi, L., Barry, D.A. 2012. Modeling of facade leaching in urban catchments. Water Resources Research 48, 12, art. no. W12503.
27. Daar S. 1994. Directory of least-toxic pest control products. Daar S. Ed. Berkeley, CA, 39pp.
28. Delaunay, T., Gourdeau, J., Hulin, A., Monteirck, S., Pernot, P., 2010. Atmospheric measurements of pesticides in France by the air quality monitoring networks. Pollution Atmospherique 208, 437-452.
29. Devault, D.A. Space-time approach of the pre-emergence herbicide contamination of the Mid-Garonne River biotope. Institute National Polytechnique de Toulouse. Toulouse, France. 2007. Ph.D thesis. 208pp. *In French.*
30. Devault, D.A., Ith, C., Merlina, G., Lim, P., Pinelli, E. 2010. Study of a vertical profile of pre-emergence herbicide contamination in middle Garonne sediments. International Journal of Environmental Analytical Chemistry 90, 3-6, 311-320

31. Diren, (2010). www.cvrh-paris.developpement-durable.gouv.fr/IMG/pdf/ Acceded July 24 2012.
32. Ecobichon D.J. 2001. Pesticide use in developing countries. Toxicology 160, 27-33.
33. Finlayson-Pitts, B.J., Pitts, J.N., 1986. Atmospheric chemistry: fundamental and experimental techniques. New York: John Wiley and Sons 638.
34. Forget, G., Goodman, T., deVilliers, A., 1993. (Eds). Impact of pesticides use on health in developing countries. Proceedings of a symposium held in Ottawa, Canada, September 17-20 1990. International Development and research Centre.
35. Fossen, M., 2006. Environmental fate of imidacloprid.. Environmental monitoring. California: Department of Pesticide Regulation.1-16.
36. Garcia-Jares, C., Regueiro, J., Barro, R., Dagnac, T., Llompart, M. (2009) Analysis of industrial contaminants in indoor air. Part 2. Emergent contaminants and pesticides. Journal of Chromatography A, 1216 (3) 567-597.
37. Gerecke, A.C., Schärer, M., Singer, H.P., Müler, S.R., Schwarznbach, R.P., Sägesser, M., Ochsenbein, U., Popow, G. 2002. sources of pesticides in surface waters in Switzerland : pesticide load through wastewater treatment plants – current situation and reduction potential. Chemosphere 48, 307-315.
38. Gilliom, R.J., 2007. Pesticides in U.S. streams and groundwater. Environmental Science & Technology, 41 (1) 3409-3414.
39. Glozier N.E., Struger J., Cessna A.J., Gledhill M., Rondeau M., Ernst W.R., Sekela M.A., Cagampan S.J., Sverko E., Murphy C., Murray J.L., Donald D.B. 2012. Occurrence of glyphosate and acidic herbicides in select urban rivers and streams in Canada, 2007. Environment Science and Pollution Research 19, 821-834.
40. Grundy, A.C., 2007. Weed occurrence on pavements in the UK: the result from a survey of Leamington Spa. Aspect of applied biology 82, Vegetation Management 175-182.
41. Gunasekara, A.S., Truong, T., Goh, K.S., Spurlock, F., Tjeerdema, R.S., 2007. Environmental fate and toxicology of fipronil. Journal of Pesticide Science 32, 189–99.

42. Hanke I., Wittmer I., Bischofberger S., Stamm C., Singer H. 2010. Relevance of urban glyphosate use for surface water quality. Chemosphere 81, 422-429.

43. Hansen, P.K., Kristoffersen, P., Kristensen K., 2004. Strategies for non-chemical weed control on public paved areas in Denmark. Pest Management Science 60, 600-604.

44. Hansson, D., Johansson, H., Kristiansson, L., Mattsson, B., 1992. Hacka rätt och ma bra –en orienterande studie om arbetsmiljön vid användingen av manuella och motordrivna redskap för ugräsbekämpning. Department of Agricultural Sciences, Swedish University of Agricultural Sciences, den. Report 154. 90pp. *In Swedish with English summary*.

45. Harnly, M.E., Bradman, A., Nishioka, M., Mckone, T.E., Smith, D., Mclaughlin, R., Kavanagh-Baird, G., R. Castorina, Eskenazi, B. 2009. Pesticides in dust from homes in an agricultural area Environmental Science and Technology 43 (23) 8767-8774.

46. Harris S.A., Solomon K.R. 1992. Human exposure to 2,4-D following controlled activities on recently sprayed turf. Journal of Environmental Science and Health 26(B), 9-22.

47. Haynes D., Müller J., Carter S. 2000. Pesticide and herbicide residues in sediment and seagrasses from the Great Barrier Reef World Heritage Area and Queensland Coast. Marine Pollution Bulletin 41, 7-12, 279-287.

48. Hein, R., 1990. The use of rotating brushes for non-chemical weed control on paved surfaces and tarmac. Department of Agriculture Engineering, Swedish University of Agricultural Sciences, Alnarp, Sweden. Report 191. 43pp. *In Swedish with English summary*.

49. Heisel, T., Schou, J., Christensen, S. & Andreasen, C. (2001). Cutting weeds with a CO2 laser, Weed Research 41: 19–29.

50. URL: http://dx.doi.org/10.1111/j.1365-3180.2001.00212. xHermanns, H., Meyer, E., Reichel, F., Bartels, U., 2006. Erprobung verschiedener Verfahreb zur Wildkrautbeseitigung in der Sportschule der Bundeswehr Warendorf (available at: http://www.wasser-und-pflanzenschutz.de/index.php?id=25). Accessed July 25 2012. *In German*.

51. Hoffman, R.S., Capel, P.D., Larson, S.J., 2000. Comparison of pesticides in eight U.S. urban streams. Environmental Toxicology and Chemistry, 19 (9) 2249–2258.
52. Jiang W., Lin K., Haver D., Qin S., Ayre G., Spurlock F., Gan J. 2010. Wash-off potential of urban use insecticides on concrete surfaces. Environmental Toxicology and Chemistry, 29, 6, 1203-1208.
53. Jiang W., Gan J. 2012. Importance of fine particles in pesticide runoff from concrete surfaces and its prediction. Environmental Science and Technology 46, 6028-6034.
54. Jorgenson, B.C., Young, T.M., 2010. Formulation effects and the off-target transport of pyrethroid insecticides from urban hard surfaces Environmental Science and Technology 44 (13) 4951-4957.
55. Jungnickel, C., Stock, F., Brandsch, T., Ranke, J. 2008. Risk assessment of biocides in roof paint: Part 1: Experimental determination and modelling of biocide leaching from roof paint. Environmental Science and Pollution Research 15, 3, 258-265.
56. Kanazawa, A., Kishi, R., 2009. Potential risk of indoor semivolatile organic compounds indoors to human health. Nippon eiseigaku zasshi. Japanese journal of hygiene 64 (3) 672-682.
57. Karickhoff S.W., Brown D.S. Scott T.A. 1979. Sorption of hydrophobic pollutants on natural sediments. Water Research 13, 241-248.
58. Karr, C., 2012. Children's environmental health in agricultural settings. Journal of Agromedecine 17 (2) 127-139.
59. Kimbrough R.A., Litke D.W., 1996. Pesticide in streams draining agricultural and urban areas in Colorado. Environmental Science Technology 30, 908-916.
60. Kreeb, F., Warnke, D., 1997. Infrarotstrahlen gegen Pflanzenbewuchs an Gleisanlangen. Einsenbahningenieur 3, 160-166. *In German*
61. Kristoffersen P., Rask A., Grundy A.C., Franzen I., Kempenaars C., Raisio J., Schroeder H., Spijker J., Verschwele A., Zarina L. 2008. A review of pesticide policies and regulations to urban amenity areas in seven European countries. Weed Research 48, 201-214.

62. Lassen, C., Skarup, S., Mikkelsen, S.H., Kjolholt, J., Nielsen, P.J., Samsoe-Petersen, L. 2001. Inventory of biocides used in Danmark. Environmental Project No 585 2001. Danish Environmental Protection Agency.

63. Lefevre, L., Blanchet, P., Angoujard, G., 2001. Non-chemical weed control in urban areas. In: The British Crop Protection Council Conference: Weeds 2001 (ed. CR. Riches), 709-714. British Crop Protection, Farnham, UK.

64. Leguay J.-P. 1999. Pollution in Middle-age. Jean-Paul Gisserot Eds. Paris. 127pp.

65. Leonard R.A., 1990. Movement of pesticides into surface waters. In Pesticides in the soil environment. Soil Science Society of America Book Series, n° 2, Madison, WI, USA, 303-349.

66. Liberman, N., Trope, Y., 1998. The role of feasibility and desirability considerations in near and distant future decisions: A test of temporal construal theory. Journal of Personality and Social Psychology, 75, 5-18.

67. Lindsey BD, Berndt MP, Katz BG, Ardis AF, Skach KA. Factors affecting water quality in selected carbonate aquifers in the United States, 1993–2005. Technical Report. United States Geological Survey. Reston, Virginia, USA; 2008.

68. Louchart, X. Voltz, M., Andrieux, P., Moussa, R., 2001. Herbicide transport to surface waters at field and watershed scales in a mediterranean vineyard area. Journal of Environmental Quality. 30 (3) 982-991.

69. Malaguerra F. Albrechtsen H.-J., Thorling L., Binning P.J. 2012. Pesticides in water supply wells in Zealand, Denmark: A statistical analysis. Science of the Total Environment. 414, 433-444.

70. Mahe C. 2007. Weeds management on roadways and green dependances in France : state-of-the-art and proposed actions. ENESAD Master Thesis 70pp.

71. Marque, F., Chabaux, P., 2006. Qualitative and economic inventory of weed control in non-crop areas. 1e conférence sur l'entretien des espaces verts, jardins, gazons, forêts, zones aquatiques et autres zones non agricoles. 15pp. *In French with English summary.*

72. Garon-Boucher/Margoum, C., 2003. Contribution à l'étude du devenir des produits phytosanitaires lors d'écoulements dans les fossés : Caractérisation physico-chimique et hydrodynamique, Thèse Cemagref de 3e cycle, Université Joseph Fourier-Grenoble. *In French.*

73. Mercadante R., Polledri E., Giavini E., Menegola E., Bertazzi P.A., Fustinoni S. 2012. Toxicology Letters 169-173.

74. Moran, K. 2010. Pesticides in urban runoff, wastewater, and surface water. Annual Urban Pesticide Use Data Report 2010. San Francisco Estuary Partnership. 38pp. www.up3project.org/documents/up3use2010_Final.pdf

75. Obendorf, S.K., Lemley, A.T., Hedge, A., Kline, A.A., Tan, K., Dokuchayeva, T., 2006. Distribution of pesticide residues within homes in central New York State. Archives of Environmental Contamination and Toxicology 50 (1) 31-44.

76. Okamura, H., Aoyama, I., Ono, Y., Nishida, T. 2003. Antifouling herbicides in the coastal waters of western Japan. Marine Pollution Bulletin 47 (1-6) , pp. 59-67.

77. Parker, C., Huntington, M., (2002). The use of herbicide application equipment for controlling weed growth on hard surfaces in amenity areas. Aspects of Applied Biology 66, 359-365.

78. Phillips, P.J., Bode, R.W., 2002, Concentrations of pesticides and pesticide degradates in the Croton River watershed in southeastern New York, July-September 2000: U.S. Geological Survey Water-Resources Investigations Report 02-4063, 20 p.

79. Phillips, P.J. and Bode, R.W., 2004, Seasonal Variability and Effects of Stormflow on Concentrations of Pesticides and their Degradates in Kisco River and Middle Branch Croton River Surface Water, Croton Reservoir System, New York, May 2000-February 2001: U.S. Geological Survey Water-Resources Investigations Report 03-4151, 16 p.

80. Pissard A., Van Bol, V., Piñeros Garcet, J.D., Harcz, P., Pussemier L. 2005. Risk indicator calculation due to pesticide use. Preliminary study : determination of practice level in Walloon Region. CERVA report. *In French.*http://mrw.wallonie.be/dgrne/eew/etudes/pesticides_CERVA_2005.pdf

81. Quiros-Alcala L., Bradman A., Nishioka M., Harnly M.E., Hubbard A., McKone T.E., Ferber J., Eskenazi B. 2011. Pesticides in house dust from urban and farmworker households in California: an observational measurement study. Environmental Health 10, 19-24.

82. Ragas, A.M.J., Oldenkamp, R., Preeker, N.L., Wernicke, J., Schlink, U., 2011. Cumulative risk assessment of chemical exposures in urban environments. Environment International 37 (5) 872-881. Rask A.M., Kristoffersen P. 2007. A review of non-chemical weed control on hard surfaces. Weed Research 47, 370-380.

83. Rask A.M. 2012. Non-chemical weed control on hard surfaces: An investigation of long-term effects of thermal weed control methods. Forest & Landscape Research No. 52-2012. Forest & Landscape. Denmark, Frederiksberg. 156 pp.

84. Revitt D.M., Ellis J.B., Lelewellyn N.R. 2002. Seasonal removal of herbicides in urban runoff. Urban Water 4, 13-19.

85. Rule, K.L., Comber, S.D.W., Ross, D., Thornton, A., Makropoulos, C.K., Rautiu, R., 2006. Sources of priority substances entering an urban wastewater catchment-trace organic chemicals. Chemosphere 63 (4) 581-591.

86. Ryberg, K.R., Vecchia, A.V., Martin, J.D., and Gilliom, R.J., 2010. Trends in pesticide concentrations in urban streams in the United States, 1992–2008: U.S. Geological Survey Scientific Investigations Report 2010–5139, 101 p.

87. Sartorato I., Zanin G., Baldoin C., De Zanche C. 2006. Observations on the potential of microwaves for weed control. Weed Research 46, 1–9

88. Schaefers, G.A., 1996. Status of Pesticide Policy and Regulations in Developing Countries. Journal of Agricultural Entomology 13 (3) 213-222.

89. Scheyer A., Morville S., Mirabel P., Millet M. 2007a. Pesticides analysed in rainwater in Alsace region (Eastern France): Comparison between urban and rural sites. Atmospheric Environment 41, 7241-7252.

90. Scheyer A., Morville S., Mirabel P., Millet M. 2007b. Variability of atmospheric pesticide concentrations between urban and rural areas during intensive pesticide application. Atmospheric Environment 41, 3604-3618.

91. Schneider, T., Sundell, J., Bischof, W., Bohgard, M., Cherrie, J. W., Clausen, P. A., Dreborg, S., Kildesø, J., Kjærgaard, S. K., Løvik, M., Pasanen, P., Skyberg, K., 2003. 'EUROPART'. Airborne particles in the indoor environment. A European interdisciplinary review of scientific evidence on associations between exposure to particles in buildings and health effects. Indoor Air 13(1) 38-48.

92. Schoknecht, U., Gruycheva, J., Mathies, H., Bergmann, H., Burkhardt, M. 2009. Leaching of biocides used in façade coatings under laboratory test conditions. Environmental Science and Technology 43 (24) , pp. 9321-9328

93. Schweinsberg, F., Abke, W., Rieth, K., Rohmann, U., Zullei-Seibert, N. 1999. Herbicide use on railway tracks for safety reasons in Germany? Toxicology Letters 107, 1-3, 201-205.

94. Shoeib, M., Harner, T., Wilford, B.H., Jones, K.C., Zhu, J., 2005. Perfluorinated sulfonamides in indoor and outdoor air and indoor dust: Occurrence, partitioning, and human exposure. Environmental Science and Technology 39, 6599-6606.

95. Schoknecht, U., Wegner, R., Horn, W., Jann, O. 2003. Emission of biocides from treated materials: Test procedures for water and air. Environmental Science and Pollution Research 10, 3, 154-161.

96. Steiner, M., Boller, M., 2004. Kupferabtrag einer Kupferfassade und Wirkamkeit der Einsenhydroxid-Kalk-Adsorberchicht zur Abtrennung von Kupfer aus dem Fassadenwasser. Eawag. *In German.*

97. Stuart M., Lapworth D., Crane E., Hart A. 2012. Review of risk from potential emerging contaminants in UK groundwater. Science of the Total Environment 416, 1-21.

98. Thuyet D.Q., Jorgenson B.C., Wissel-Tyson C., Wanatabe H., Young T.M. 2012. Science of the Total Environment 414, 515-524.

99. Trauth R., Xanthopoulos C. 1997. Non-point pollution of groundwater in urban areas. Water Research 31, 11, 2711-2718.

100. Ulén, B., Kreuger, J., Sundin, P., 2002. Undersökning av beka°mpningsmedel i vatten fra˚n jordbruk och samhälle år 2001. Report 2002:4. Ekohydrologi 63. *In Swedish*

101. US Environmental Protection Agency, 2008. Reregistration Eligibility Decision for Prometon. Office of Pesticide Programs. 63pp. http://www.epa.gov/oppsrrd1/REDs/prometon-red.pdf

102. Van de Voorde, 2012. Incidence des pratiques d'entretien des toitures sur la qualité des eaux de ruissellement. Cas des traitements par produits biocides. Ph.D. thesis. Université Paris Est, Paris, France. 277pp. *In French.*

103. Van Maele-Fabry, G., Lantin, A.-C., Hoet, P., Lison, D., 2011. Residential exposure to pesticides and childhood leukaemia: A systematic review and meta-analysis. Environment International 37 (1) 280-291.

104. Walters, J., Goh, K.S., Li, L., Feng, H., Hernandez, J., White, J., 2003. Environmental monitoring of carbaryl applied in urban areas to control the glassy-winged sharpshooter in California. Environmental Monitoring and Assessment 82 (3) 265-280.

105. Webster, D., 2000. Malaria kills one child every 30 seconds. Smithsonian 31 (6) 33-44.

106. Weschler, C.J., Nazaroff, W.W., 2008. Semivolatile organic compounds in indoor environments. Atmospheric Environment 42, 9018-9040.

107. Weston, D.P., Lydy, M.J. 2012. Stormwater input of pyrethroid insecticides to an urban river. Environmental Toxicology and Chemistry 31 (7) , pp. 1579-1586

108. Whitmore, R.W., Kelly, J.E., Reading, P.L., 1992. National Home and Garden Pesticide Use Survey: Final Report. US EPA, Office of Pesticides and Toxic Substances, Biological and Economic Analysis Branch. Research Triangle Institute. Vol. 1. NTIS PB92-174739.

109. Wittmer I.K., Bader H.-P., Scheidegger R., Singer H., Lück A., Hanke I., Carlsson C., Stamm C. 2010. Significance of urban and agricultural land use for biocide and pesticide dynamics in surface waters. Water Research 44, 2850-2862.

110. Wittmerr, I.K., Scheidegger, R., Urban biocide Stamm, C., Gujer, W., Bader, H.-P., 2011a. Modelling biocide leaching from facades. Water Research 45 (11) 3453-3460.

111. Wittmer, I.K., Singer, H., Scheidegger, R., Bader, H.P., Stamm, C., 2011b. Loss rates of urban biocides can exceed agricultural pesticide loss rates. Science of the Total Environment 409, 920-932.

112. Wood, R., 2004. Urban weed control: Innovations in kerb and channel weed management. In: 14th Australian Weeds Conference, Wagga-Wagga, New South Wales, Australia, 6-9-September- (eds B Sindel & S Johnson), 210–211. RG and FG Richardson, Melbourne, Sydney, Australia.

113. Zadjian, E. 2004. Nuisances des mauvaises herbes et propositions de seuils d'intervention pour le désherbage en zone urbaine. Mémoire de fin d'études. INH. *In French*

Chapter 5

Like a Canary in the Coal Mine: Behavioral Change as an Early Warning Sign of Neurotoxicological Damage

Kathleen M. Raley-Susman[1]

[1]Department of Biology, Vassar College, Poughkeepsie, NY, USA

INTRODUCTION

With more than 7 billion people in the world, global food production is of critical importance. Back in the 1960's, when population models predicted such a large human population, agricultural scientists and policy makers felt an urgent need to dramatically increase global food production and to radically change how food was grown and distributed. Science, business and governments collaborated to engineer a paradigm shift in how we feed our species. Galvanized by the efforts of Norman Borlaug to successfully develop high-yield and disease-resistant wheat, the Green Revolution led to an order of magnitude increase in global food production in fewer than forty years [1]. Both the Rockefeller and Ford Foundations worked with international governments to use a combination of agricultural and technological approaches to vastly increase the productivity of the major cereal crops that feed the world: wheat, corn, rice, and later other important crops.

The major changes in agricultural practice included the selective breeding of high yield crop varieties, changes to irrigation approaches, increased mechanization of farms and the use of fertilizers and pesticides. The high-yield varieties were selectively bred to produce large and many seeds and to have shorter and sturdier stems to support the seeds. They matured quickly and were much less sensitive to photoperiod, enabling a longer growing season or multiple growing seasons in a year. They were also bred to respond to externally applied fertilizers with fast growth and rapid maturation [1]. Farmers were eager to plant the super-crops and soon changed agricultural approaches to focus exclusively on a few or one variety, giving rise to the monoculture agriculture and large mechanized farming that dominates global agricultural practice today, particularly in the developed world. These varieties are dependent on plentiful irrigation and fertilizer application. To protect the new crops and further enhance yield, farmers applied a growing arsenal of chemical pesticides, some of them derived from chemicals produced in World War I and World War II as chemical warfare agents.

The Green Revolution prevented countless deaths due to starvation and created jobs and entire new industries. Huge mega-companies devoted to the production of pesticides, fertilizers and seeds now dominate big agri-business. More than one third of the world's workers are employed in agriculture. The costs are also enormous. Industrial agriculture has caused widespread damage to the land, water and air. Fertilizer run-off has damaged aquatic ecosystems; soils overplanted with monoculture crops are depleted of nutrients; soil, water and the air are contaminated with a complex mixture of pesticides and other toxic chemicals. What are the consequences of chronic exposure to low levels of mixtures of pesticides in the water we drink, in the air we breathe? How are other organisms affected? How are beneficial soil organisms influenced? There is growing awareness of the human health and environmental health consequences of our high-yield agricultural practices. An increasing number of studies indicate that most animals are affected deleteriously by the pervasive presence of pesticides in the environment, humans included.

This chapter focuses on the effects of pesticides and pesticide mixtures on unintended organisms. I explore the behavioral consequences of exposure to toxic chemicals in the environment, with a focus on the use of behavioral change as an early marker of potential or ongoing damage to the nervous system. Recent work in model organisms underscores the value of behavioral assessment of toxicity. Because of the substantial evolutionary conservation of the molecular targets of many pesticide chemicals, what we learn about in non-human animals can readily be applied to human health. We are all in this together.

THE MOST COMMON PESTICIDES

The hundreds of pesticides in existence fall into five main categories: organophosphates, organochlorines, carbamates, glyphosates and neonicotinoids. Agricultural workers are often exposed to high doses, particularly over their lifetime. In addition, most rivers and lakes, non-agricultural soil and even the air contain measurable, albeit lower, concentrations of all these chemicals. The vast majority of pesticides in use to protect plants from herbivory target the control of muscle contraction, leading to paralysis and death of the intended target insects and other arthropods. Fungicides and herbicides target enzyme pathways that result in the death of harmful fungi and weeds, but the chemicals can have unwitting effects on animals as well. In order to better understand how these pesticides operate in animals, I will first briefly review the mechanism by which muscles are activated by nerves.

A Key Target of Pesticides: The Neuromuscular Junction

The connection between motor neurons that generate movement and the skeletal muscles they activate is called the neuromuscular junction (NMJ). This synapse is highly specialized to afford rapid and reliable activation of the muscle. In vertebrates, all skeletal muscles are activated by one type of neuron, the motor neuron, which uses the neurotransmitter acetylcholine (Figure 1). The presynaptic terminal of the neuromuscular junction contains hundreds of synaptic vesicles, each loaded with 5000- 10,000 molecules of acetylcholine. A single action potential firing from one motor neuron can cause the release of as many as 300 vesicles within a millisecond or so [2]. The 1.5 million molecules of transmitter rapidly diffuse across the cleft between the nerve and muscle cell and interact with 1000-2000 receptor channels that are densely clustered at the synaptic junction membrane [3], leading to a large depolarization that initiates a muscle contraction. The cytoarchitecture of this synaptic connection has a tremendously large safety factor, to ensure a 1:1 correspondence between an action potential in the nerve and a corresponding action potential (leading to a contraction cycle) in the muscle cell.

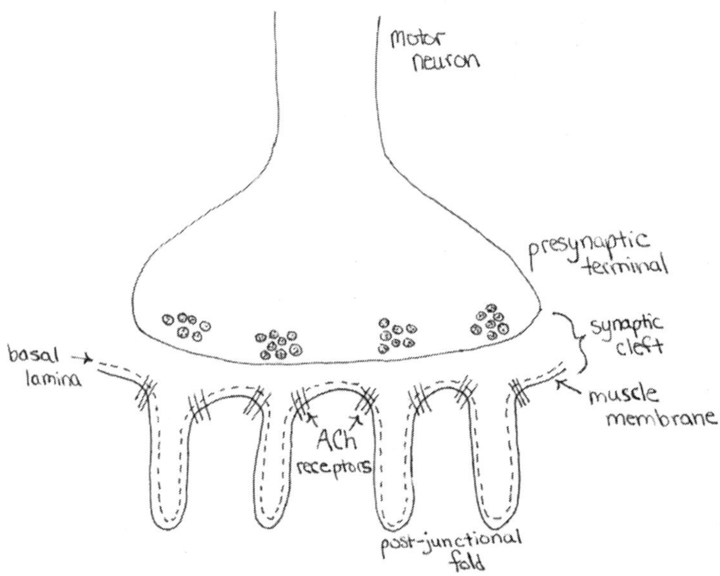

Figure 1: Vertebrate Neuromuscular Junction.

At the vertebrate neuromuscular junction, the presynaptic nerve membrane and the postsynaptic muscle membrane are separated by a small space called the synaptic cleft that also contains an extracellular matrix-derived basal lamina that adheres tightly to the muscle membrane (Figure 1). Composed of structural molecules like collagen type –IV and laminins [4], the basal lamina also contains a form of acetylcholinesterase with a collagen tail (ColQ-AChE). The ColQ tail localizes this form of acetylcholinesterase to the NMJ [5].

Acetylcholinesterase is the product of a single, highly conserved vertebrate gene, and many different versions are generated by alternative splicing, co-translational and post-translational modifications and even the formation of oligomers [6], differing in tissue expression and subcellular localization. The catalytic region of the enzyme is very highly conserved across organisms. As a consequence, chemicals that generally affect cholinesterases can have effects on multiple tissues in multiple organisms.

There are hundreds of different proteins involved in the molecular mechanism of neurotransmitter release and the majority of these proteins are involved in the release of all neurotransmitters. The proteins govern the movement of synaptic vesicles to the presynaptic terminal membrane, the fusion of the vesicles with the membrane so as to release the chemical neurotransmitter, the recycling of the vesicle membranes back into the terminal and the packaging of neurotransmitter molecules back into vesicles. These important molecular players are highly conserved evolutionarily, being present in mammals, amphibians, flies and nematodes. In addition, all vertebrates utilize the same overall scheme for muscle contraction. Many of these proteins are found in all animals, including invertebrates. Nonetheless, there can be substantial genetic variation in the proteins, with differences in function and structure in vertebrate as compared with invertebrate neurotransmission proteins. The evolutionary divergence of these proteins is important in the design of pesticides that are selective for invertebrate target species, while at the same time being relatively less toxic to non-target animals. However, many of the pesticides currently in use do have considerable cross-reactivity with non-target organisms, depending on the dose and exposure methods used.

Invertebrates have an overall similar mechanism by which motor neurons excite muscles and many of the same proteins are critical for the release of neurotransmitter. There are a number of differences, however. While acetylcholine is the major excitatory neurotransmitter used by motor neurons in a large number of invertebrates including nematodes, annelids, arachnids and mollusks, insects and other arthropods utilize glutamate as a major excitatory neurotransmitter at the neuromuscular junction, with glutamate acting on postsynaptic muscle membrane glutamatergic sodium channels. Another key difference is that many invertebrates also control muscle movement via inhibitory motor neurons that utilize GABA or glycine as a neurotransmitter.

These inhibitory motor neurons synapse directly on muscle fibers [7]. This local control of muscle function is quite different from the central spinal cord-level control in vertebrates [8]. Another distinction between vertebrate and invertebrate muscle control is that arthropods in particular modulate muscle function peripherally, rather than at the level of the central nervous system, via a wide variety of circulating hormones and neuromodulators [7]. Many invertebrate motor neurons release more than one neurotransmitter, achieving modulation of muscle force directly at the neuromuscular junction [7]. While ACh is an important neurotransmitter across all animals, the principle effect of pesticides directed at the acetylcholine neurotransmitter system is to alter central nervous function in arthropods.

Organophosphates

More than half of the insecticides in use across the globe fall into the organophosphorous group of chemicals. The major organophosphates in use today are chlorpyrifos, parathion, malathion and diazinon. Chlorpyrifos is the major compound used in the United States and Europe. Organophosphates were first developed as nerve gases in the 1930's in Nazi Germany [9]. After World War II, American chemists developed this class of insecticide based on those materials. It is estimated that more than a billion pounds are produced worldwide each year [9].

Organophosphates target the enzyme acetylcholinesterase, which is a member of the serine hydrolase enzyme superfamily [10]. This enzyme plays a critical role in nerve and muscle activity in vertebrates and some invertebrates (Fig. 1), and in central nervous system function in many animals, including arthropods. The organic groups of organophosphate pesticides can be modified to affect target specificity, penetration of the chemical internally, water solubility and persistence in the environment. All of the major organophosphorous pesticides in use for agricultural and residence purposes have a common mechanism of action, that of phosphorylating acetylcholinesterase, thereby inactivating it [11]. This covalent modification is permanent. The inhibition of acetylcholinesterase activity, particularly at the neuromuscular junction, reduces the clearance of acetylcholine from the synaptic, resulting in prolonged neuromuscular stimulation, which causes seizures and paralysis. In arthropods, the effect is primarily within central regulatory neurons and sensory systems [12].

Unintended Targets of Organophosphorus Pesticides

When a pesticide is developed for use, it must be approved by governmental regulatory agencies in many countries. The Environmental Protection Agency (EPA) in the United States requires testing of toxicity to a variety of animals, usually assessing lethality at working doses recommended by the pesticide-producing company, as well as assessing potential cancer risk. The EPA requires pesticides be assessed for ecological as well as human health risks. A company seeking to gain EPA approval to sell and distribute a pesticide conducts the scientific tests and files the regulatory materials. Pesticides that have gained approval are periodically reassessed based on reports received from scientific study of risk and reports of problems submitted to the EPA. Ecological assessment data include toxicological studies on wildlife and plants that represent non-target organisms likely to be exposed unintentionally through runoff, aerial drift or bioaccumulation. For this testing, organisms are exposed to different concentrations of the "active ingredient" of the pesticide being assessed, usually in isolation, without other so-called inactive ingredients. Both short and long-term effects of the active ingredient are measured including lethality, growth and reproduction rates. In addition, residue measurements are conducted to determine degradation of the pesticide, possible toxicity of breakdown metabolites, persistence and ability to travel in the environment (soil, water, air, bioaccumulation). If the regulatory commission (like the EPA) believes that a particular type of ecosystem is not likely to be exposed to a candidate pesticide, then the registration materials stipulate that the pesticide is either "safe" or poses no "reasonable harm" to the organisms within that ecosystem, even if direct evaluation of risk or exposure has not been conducted. If the assessors predict a high likelihood a particular ecosystem will be exposed, then additional tests are often required [13].

Acetylcholinesterase, the major target enzyme for organophosphorus pesticides, is a member of a very large class of carboxylic esterases [14]. Most organophosphates interact with a serine group present at the active site of the enzyme. However, they can also affect other enzymes, including other serine hydrolases and even serine proteases

[15]. As a result, many cell processes and cell signaling pathways can be disrupted, including serotonin and dopamine neurotransmission, growth regulation, hormonal regulation and other systems. Of particular concern to epidemiologists are the possible long-term consequences of exposure during key developmental critical periods. For example, 83% of pregnant women in the U.S. had detectable levels of an organophosphate metabolite in their urine [16]. At the same time, a 600% increase in autism incidence has been reported in California [17]. Mothers living in the California Central Valley who had been exposed to pesticides while pregnant gave birth to children who were 7.6 times more likely to be diagnosed with Autism Spectrum Disorder [18] and there was a 230% increase in maternally reported pervasive developmental disorders in children whose mothers had measurable organophosphate metabolites in their urine [19]. Cholinergic abnormalities have been reported in autism [20]. While not conclusive, studies such as these and others [20] indicate that gestational exposure to pesticides might well influence neural development. Direct experimental animal studies support these concerns. Exposure to chlorpyrifos during the early postnatal period altered rat memory function and spatial navigation in adulthood [21]. Chick embryos exposed to low levels of chlorpyrifos showed reduced head development [22] and axon development was disrupted in exposed zebra fish embryos [23].

Acetylcholinesterase (AChE) inhibition, affecting neural cholinergic signaling during development, is a likely cause of organophosphorus pesticide effects on behavior, growth, and reproduction. However, most organophosphorus pesticides can also affect other enzyme systems that involve serine esterases, serine hydrolases or serine proteases. This large and essential group of enzymes is important for overall cell and body metabolism, immune and endocrine system functioning, in addition to nervous system functioning. These enzymes differ in their sensitivities to different organophosphorus pesticides, from essentially insensitive to highly sensitive to inhibition. Doses of organophosphate compounds that exquisitely affect AChE activity might have little or no effect on other members of this enzyme superfamily. However, many unintended target enzymes are affected by concentrations of organophosphate compounds or their metabolites that have little effect on AChE. For example, several organophosphates that induce a delayed neuropathy are relatively less reactive with AChE [24]. Further, metabolites of organophosphorus insecticides can also exert effects by reacting with non-target members of the serine hydrolase enzyme superfamily [25]. Indeed, such interactions with serine hydrolases in liver are important means for detoxification [25]. Recent work on the environmentally relevant metabolites of chlorpyrifos and parathion demonstrated substantial inhibition of the activity of key liver carboxylesterases [25]. Several lipases important in brain function are sensitive to organophosphates. Neuropathy target esterase (NTE) is one such enzyme, first discovered as the target enzyme associated with a lethal neuropathy caused by a ginger extract substitute added to drinks during Prohibition in the 1930's [10], an insecticide that was also used as a lubricant for machine parts.

Over the past 15 years or so since the introduction of organophosphorus pesticides as a safer alternative to DDT and other organochlorine pesticides, increasing evidence indicates that organophosphates are harmful to immune function through both acetylcholinesterase inhibition and via noncholinergic mechanisms. Vertebrates and invertebrates share common cell types and mechanisms of innate immunity, namely the presence of phagocytic cells (monocytes and neutrophils), the production of cytokines [26] and a histocompatibility cell recognition system. The complex adaptive immune responses, in particular T- and B-lymphocyte antibody-mediated immune systems first appeared with the evolution of the vertebrates. Serine hydrolases play important roles in innate immune function and could be unintended targets of organophosphates [27]. Malathion, at doses that inhibit acetylcholinesterase, suppressed humoral immune responses in mice exposed under laboratory conditions [28] and inhibited cytotoxic T-lymphocyte responses [29]. At low doses, below those that inhibit AChE, malathion stimulated immune activation in rodents [30]. Immunotoxic effects have also been reported in birds, fish, small mammals and soil invertebrates exposed to chlorpyrifos, malathion and diazinon [26]. Parathion, diazinon and chlorpyrifos have also each been shown to alter immune function [26]. Reports on earthworms have shown reduced immune function after exposure to organophosphorus compounds [31, 32]. Laboratory studies in fish also demonstrated an immunosuppression after sublethal organophosphate exposure to malathion [33] and diazinon [34]. A number of transcription factors important for cell division and differentiation are inhibited by chlorpyrifos [35]. These studies are complex and can suffer from interpretational difficulties and clearly much more work is needed. Nonetheless, the increasing reports of unintended effects on many different organ systems are cause for concern.

Organophosphorus Pesticides Act as Endocrine Disrupters (EDC'S)

An endocrine disrupting chemical (EDC) is any chemical that alters an endocrine-regulated system or behavior. Early work centered on chemicals whose structural characteristics allowed them to be estrogen-mimics or estrogen receptor agonists or antagonists. However, more recent work has acknowledged that many chemicals, while not directly interacting with estrogen receptors, can alter endocrine-related functions and so produce long-term effects on growth, metabolism, reproduction and behavior [36].

Developmental exposure to organophosphorus pesticides has widespread effects on neural development, which influences behavior and physiology well into adulthood. Late pre-natal and early postnatal exposure of rats to low concentrations of chlorpyrifos caused impairments in synaptic function, with subsequent locomotory and cognitive dysfunction ([37- 39]). Exposure to very low levels of chlorpyrifos early in gestation, at the time of neurulation, a very early stage of brain development, resulted in locomotory and cognitive abnormalities that persisted into adulthood [40]. The impairments were not severe, suggesting some neurodevelopmental compensation or recruitment of alternate mechanisms to preserve functionality. Indeed, early developmental exposures to chlorpyrifos, parathion, malathion or diazinon have each been associated with alterations in serotonergic and dopaminergic systems [41]. The consequences of alterations in neurotransmitter systems during development, referred to as organizational effects, would have an impact on neuroendocrine system development, thus altering reproductive behavior, in addition to myriad effects on sensory function, cognitive function, predator/ prey avoidance and other behaviors. For example, salmon exposed to diazinon were less responsive to predator cues [42]. Organophosphate exposure in development has also been linked to modified oxytocin and arginine vasopressin (AVP) activity [43], altered social behavior [44] and sex-specific behaviors [45].

There is sufficient evidence that developmental exposure to organophosphorus pesticides can cause permanent neurobehavioral and neurological impairments that chlorpyrifos has been banned in the US for residential use. Despite this compelling evidence, however, this and other organophosphorus pesticides are still heavily used in agricultural settings, exposing workers, nearby communities and the surrounding aquatic and terrestrial wildlife to these harmful and environmentally persistent chemicals. It is abundantly clear that, while recommended application doses of particular organophosphorus pesticides may not cause overt death, persistent exposure to these compounds, present at biologically active concentrations in water, soil and the food supply, can lead to neurobehavioral changes that influence wildlife and humans.

Organochlorine Pesticides

Organochlorine pesticides are among the most environmentally persistent human pollutants, existing in soil, water and air for decades. Developed in 1946, the first commercial herbicides, including 2,4 D and 2,4,5, T were introduced to control broad-leaf weeds. 2,4,5 T was used extensively during the Vietnam war to clear jungle hiding places. Byproducts of that chemical production included dioxin and DDT, which are potent neurotoxins. In the 1940's, DDT was heralded as a potential eradicator of malaria and was sprayed widely as a residential and agricultural insecticide until alarming reports of bird and human morbidity and death in the 1960's. Other banned organochlorine pesticides include aldrin, dieldrin, chlordane and heptachlor. Despite recent bans on their use in some countries, measurable levels of many of these chemicals are still present, even in areas like the Arctic, where they were never used [46].

Others remain in use in the US and in the developing world, including lindane, endosulfan, dicofol, methoxychlor and pentachlorophenol. These chlorinated hydrocarbons are readily lipid soluble, which leads to bioaccumulation in wildlife and in the food chain. Measurable organochlorine residues are found in blood, adipose and breastmilk in humans. Most organochlorines interfere with ion channel function, predominantly in the nervous system. For example, endosulfan binds to and blocks the chloride channel portion of the GABA-a receptor, acting as a noncompetitive antagonist [47]. Lindane, an active ingredient in head lice treatments for children, also interacts with the GABA-a receptor, whereas methoxychlor, chemically similar to DDT, interacts with insect sodium channels, leading to their persistent activation [48]. Lindane and several other organochlorines have been banned from use in many countries because of unintended endocrine-disrupting effects on wildlife and humans (see below) but others, including endosulfan, continue to be used heavily in many parts of the world.

The primary target of endosulfan as an insecticide is to alter ion flux of sodium and potassium, thereby interfering with neuron activity and thus motor function in insects. However, it is a relatively nonspecific insecticide, which is in part why it is used widely for pest protection of many important crops worldwide, as well as for a preservation treatment for wood. Endosulfan, known to be toxic to fish and other aquatic organisms [47], blocks the GABA receptor ion channels, interfering with critical inhibitory neuronal function in vertebrates, resulting in hyperexcitability of many neuronal circuits. Endosulfan is highly toxic to fish and aquatic invertebrates [49] and has also been found to be genotoxic and neurotoxic to mammals [50]. Endosulfan can be transported by air and water and measurable and increasing levels have been reported in the air over the Arctic [50] and even within tissues of animals living large distances from regions that use pesticides, like polar bears in the Arctic [46]. As a result, endosulfan used in one nation will affect many other nations that might have banned its use. Organochlorines are found globally and in every ecosystem that has been studied [36].

Unintended Targets of Organochlorines

Organochlorines, like organophosphates, have a number of unintended effects, the nature of which varies with exposure dose/time, species, developmental stage and method of exposure. Chronic sublethal exposure of adult rats to endosulfan in food led to hair loss, enlarged kidneys and increased liver toxicity. Higher doses led to increased incidence of aneurysms [47]. Dogs exposed to endosulfan exhibited increased sensitivity to noise, jerky or tonic movements, excessive abdominal contractions and neurotoxicity [47]. These effects are not related to their intended effects on GABA-a channels and occur at lower concentrations.

In addition to the acute and pervasive neurotoxicity of the major organochlorine pesticides, the compounds in lower concentrations act as endocrine disrupting chemicals (EDC's). Low levels of endosulfan impaired sexual pheromone communication and mating success in newts (*Notophthalmus viridescens*) [51] and methoxychlor affected scent-marking behavior in mice [52]. Further, mice exposed to methoxychlor *in utero* exhibited altered exploratory behavior when adults [53]. Frogs exposed to endosulfan exhibited convulsions, followed by temporary paralysis [54]. Rats exposed to endosulfan as adults or early in development exhibit an array of neurological alterations, including impairments in learning and memory, increased spontaneous motor activity, reduced escape and avoidance behavior learning and increased fighting behavior [47]. Hormones modulate these behaviors, particularly during neural development. Thus, while endosulfan is not a direct endocrine disrupter [47], its potent effects on neural systems may indirectly influence many endocrine-regulated functions.

Carbamates

Carbamates, introduced in the 1950's, are organic insecticides and fungicides that were originally derived from the extracts of the West African calabar bean. This class of pesticide inhibits acetylcholinesterase activity much like organophosphate pesticides, but the effects are reversible. Carbamates degrade fairly rapidly in water, with half-lives ranging from 1 day to 4 weeks. The liver detoxifies carbamates and the metabolites are excreted via the kidneys. The major insecticides are aldicarb, carbofuran (Furadan) and carbaryl (Sevin). Carbamates are also found in cosmetics, as preservatives and in polyurethanes, so their presence in the home is quite pervasive [55]. Humans are exposed via skin absorption, ingestion and inhalation and acute toxicity symptoms include headache, nausea, and dizziness. Carbamates are the class of insecticides most prevalent for in-home use. Dithiocarbamates, in contrast, while still rapidly degraded, have little to no effect on acetylcholinesterase activity but rather bind divalent metal ions like manganese and zinc and are used as potent fungicides for home and agricultural use.

Carbamates, despite their rapid degradation and overall low persistence and mobility within the environment, are highly toxic to a wide range of vertebrates and invertebrates. Insecticidal carbamates cause the same symptoms of neurotoxicity as the organophosphate pesticides, with a similar wide range of LD50 values based on chemical modifications [55]. Because the chemicals have low persistence and low mobility, and because the actions are reversible, carbamates are generally considered safe for humans and wildlife, except for occupational or intentional exposure to high concentrations. Very little work on possible long-term consequences of brief or transient exposure to the chemicals has been done, nor has there been much work on possible genotoxic, carcinogenic or reproductive effects. A recent review of the available epidemiological studies of carbamates indicated variable cancer risk from occupational exposure to different carbamates [56]. A few reports indicate reproductive and developmental problems in children of exposed workers [57, 58], suggesting possible trans-generational effects.

There is growing concern about carbamates, particularly thiocarbamates, and in particular the associated metal toxicity. A recent human epidemiological study implicated the carbamate fungicide ziram in the etiology of Parkinson's disease [59]. The risk of Parkinson's disease was significantly higher when patients were exposed to the combination of ziram, maneb (another carbamate fungicide) and paraquat, for patients exposed in both occupational and residential areas, and for patients exposed at a younger age [59]. The manganese associated with several common dithiocarbamate fungicides is released when the organic carbamate moiety is degraded or metabolized which can enhance the toxicity of these compounds [60]. Manganese is a potent neurotoxin that is particularly harmful to dopamine neurons [61], causing toxicity via oxidative stress, mitochondrial inhibition and the production of reactive oxygen radicals. Dopamine neurons are particularly vulnerable because both extracellular and intracellular dopamine is enzymatically oxidized to a reactive species in the presence of manganese [61]. Mancozeb exposure, while not lethal to nematodes, inhibited larval growth, induced a heat shock response and altered gene expression [60]. These results suggest that, while dithiocarbamate fungicide might not be lethal or toxic in the short-term, it might cause long-term changes in neuron function that influence behavior and health later in life.

Glyphosate (Roundup)

Introduced by Monsanto in the 1970's, glyphosate is one of the most prevalent herbicides in use today. In 2007, the US applied 185 million pounds of Roundup for agricultural use alone [62]. Monsanto also developed glyphosate-resistant crops, so many farmers plant these crops and apply copious amounts of Roundup. It is also the herbicide most used residentially and on corporate campuses for lawn and garden maintenance.

The primary mechanism of action of glyphosate is to inhibit the enzyme 5-enolpyruvylshikimic acid-3- phosphate synthase, an enzyme found only in plants and microorganisms that is responsible for the synthesis of tyrosine, tryptophan and phenylalanine [63]. Because animals obtain these amino acids from dietary sources, glyphosate has been heralded as a safe and effective herbicide for general use. Glyphosate is considered non-carcinogenic and of very low toxicity to humans. It is less persistent in water than in soil, where it can be retained for over a year. Glyphosate has been considered only slightly toxic to amphibians and fish because of its 12-60 day persistence in water [64].

With its widespread use, glyphosate levels are measurable in most soil, water and even air. Low levels have been shown to inhibit steroidogenesis [65] and to alter testosterone levels and testicular morphology in rats exposed pre-pubescently to glyphosate [66]. There are also reports that the commercial formulations of Roundup, which contain inactive ingredients like surfactants to enhance persistence and ease of application, have teratogenic effects in amphibians and other vertebrates [67, 68]. Several studies have demonstrated that Roundup inhibits aromatase, which can have profound neuroendocrine effects [69], although these findings are controversial because these effects have so far only been demonstrated in isolated human cells. A recent report provided evidence that male offspring of female rats exposed to concentrations of Roundup in their drinking water that are found in the environment exhibited alterations in genes important for thyroid function [66]. These reports, taken together, suggest that glyphosate, or the combination of glyphosate and its inactive ingredients in commercial formulations, may cause changes in neural and neuroendocrine function that could have long-term consequences.

Neonicotinoids

In the 1980's and 1990's, largely in response to the toxicity caused by organophosphate and organochlorine pesticides, a new class of insecticide was developed based on the naturally occurring plant alkaloid nicotine, which is known to have a higher toxicity in insects than in mammals. Bayer Corporation introduced imidacloprid in the 1990's and it is now the most widely used neonicotinoid insecticide in the world [70]. Imidacloprid and two other prevalent neonicotinoids, clothianidin and thiamethoxam, are nicotine-related compounds that are photostable, water-soluble and environmentally persistent. Because they act selectively on the acetylcholine receptor, much like nicotine (Figure 1), it is thought that there are far fewer unintended targets for the compounds. Neonicotinoids are not metabolized by acetylcholinesterase and their clearance from tissue is slow and so their action on the acetylcholine receptor is essentially irreversible. The neonicotinoid insecticides have a much higher affinity for the insect nicotinic acetylcholine receptor than for other such receptors, leading to a lower toxicity in non-insect animals [71].

Neonicotinoids are of deep concern because they act with high potency against target insect pests as well as beneficial insects like honey bees and other important pollinators. A host of recent work has indicated that low-level exposures, of the magnitude measured in agricultural dust drift, pollen and nectar, impairs honey bee navigation, foraging and queen production [72], particularly when exposures occur alongside other pesticides [73].

Even though neonicotinoids are not lethal to non–insects, low, sublethal exposures impaired feeding behavior in a crucial benthic freshwater invertebrate, *Gammarus pulex* [74], and reduced exploratory and burrowing behavior in two different species of earthworm [75]. Further, male offspring of female mice exposed to chothianidin during gestation and lactation had long-term alterations in exploratory behavior and increased spontaneous locomotory activity [76]. These findings suggest that neonicotinoids are not safe for non-target organisms.

ASSESSING BEHAVIORAL CHANGE RATHER THAN LETHALITY

Determining whether a pesticide is harmful to humans or wildlife is a tricky business. Many different factors are important. How the chemical spreads through water, soil, and air affects the extent of pesticide exposure. How the chemical is applied (spraying, irrigation water, seed coating) also affects how it spreads through the environment. Temperature, pH, sunlight and other environmental conditions affect the persistence or breakdown of the chemical. The biological activity of the active ingredient can be altered by other so-called inert ingredients contained within the commercial formulation, like surfactants or solubilizers. The breakdown products might also carry risks in and of themselves. The action of a particular compound might also be different if it is present in a mixture of other pesticides. The concentration of the compound can also have different effects, both on different organisms and on different biological processes within the same organism. The time of exposure (long-term, chronic, repeated) as well as the time of life (early in development, in adulthood) also influence the kind or nature or extent of the biological effects. It is not surprising, then, that regulatory agencies focus on overt, direct toxicity or lethality of the active ingredient of a mixture. It is also not surprising that the focus is on major aspects like carcinogenicity or acute toxicity in humans and animals likely to encounter high concentrations of the compounds being tested for use.

Because so many pesticides have unintended effects and long-term consequences, it makes sense to develop assays of nonlethal effects on reproduction, behavior and nervous system function of exposure to field-levels of the compounds. A rational approach being taken by an increasing number of investigators is to develop behavioral assays in a number of model organisms that have proven to be useful in testing the efficacy of pharmaceutical compounds and that span animal taxa. Understanding the evolutionary similarities among these organisms can provide insight into not only the choice of the model organism, but also the types of behavioral assays that would be most useful. Further, appreciating the differences in mechanism of action or mode of interaction of particular pesticides across evolutionarily distinct taxa will also help inform toxicological testing approaches. While many of the current strategies try to take these kinds of approaches, the continued emphasis on lethality measures, particularly in the licensing approval process, quite often underestimates risks that might lead to long-term morbidity in humans and other animals.

Efforts to explore behavioral effects of pesticides and mixtures of pesticides have been quite successful. As early as 1958, Broley observed abnormal nesting, courtship and reproductive in the Florida bald eagle in areas that were sprayed with DDT [77]. Guppies exposed to the fungicide vinclozolin exhibited changes in body coloration and courtship behavior [78]. The courtship behavior of a number of different bird species, including ringed turtle doves, Japanese quail and Western gulls was adversely affected by exposure to DDT [36]. Many experimental studies in rodents have demonstrated changes in reproductive behavior, locomotory behavior and motivational behavior in response to pesticide exposure [36]. Organophosphates, organochlorines and carbamate pesticides have been shown to alter chemosensory behavior in salmon and mice [52, 79], social behavior in fish [80] and mice [81] and behaviors in tadpoles [82], birds [83] and mammals [84].

Behavioral changes can be more sensitive than other measures, like lethality or neurotoxicity. For example, swimming behavior in fish was more sensitive to toxic stress than either lethality or growth [85]. With the growing recognition that behavioral assays can be effective ways to determine possible sublethal and long-term effects of various pesticides, model organisms like the fruit fly *Drosophila melanogaster* are proving to be effective screens for pesticide sensitivity. Of course, because *Drosophila*species are insects, they may not be the best models for evaluating effects of pesticides, particularly insecticides, on unintended target organisms. But studies in closely related organisms are crucial for understanding the effects of pesticides on beneficial organisms, like honey bees and other insect pollinators that are also unintended targets of pesticides. *Drosophila melanogaster* has been historically used for evaluating genotoxicity of pesticides and heavy metals [86]. A number of straightforward behavioral assays, like locomotion, courtship behavior, and aggressive behavior, have proven quite useful in assessing the neurotoxicity and cellular mechanisms of toxicity of various chemicals.

Caenorhabditis Elegans Behavior

Caenorhabditis elegans (C. elegans), a free living soil nematode with a short life cycle, high fecundity and ease of maintenance in a laboratory, is a valuable model organism to use in neurotoxicology, genotoxicity or behavioral toxicology measures. Its genetic tractability also enables mechanistic studies. The *C. elegans* genome contains 60-80% of the genes found in humans and shares many of the same mechanisms of neuron function, development, gene regulation and signal transduction pathways [87]. Early work with this organism examined behavioral toxicity of heavy metals like copper, lead and mercury. Locomotory behaviors like head movement, feeding behavior and simple learning behaviors all showed sensitivity to heavy metal exposure [87- 89]. The herbicide paraquat (methyl viologen) has been extensively studied using *C. elegans*, particularly in response to emerging epidemiological evidence for an increased incidence of Parkinson's disease in humans exposed to paraquat [90]. Using *C. elegans*, several potential mechanisms involved in paraquat dopamine neurotoxicity have been described, including oxidative stress leading to reactive oxygen species production [91].

A growing number of studies are accumulating on the adverse behavioral effects on reproduction behaviors, general locomotion or feeding behaviors of sublethal exposure to various pesticides including aldicarb [92], juglone and paraquat [93, 94], organophosphorous compounds [95, 96] and carbamates [97, 61, 98]. The LD50 values for these pesticides are similar to those in rats and mice, further validating this model organism for studies of non-target animal toxicity. Sublethal, environmentally relevant concentrations of all of these different environmental contaminants have substantial effects on behavioral processes. Two organophosphorous insecticides, malathion and vapona, inhibit locomotion at sublethal concentrations [99]. Aldicarb, a potent acetylcholinesterase inhibitor, causes paralysis in *C. elegans* [92], likely because acetylcholine is the excitatory transmitter used at *C. elegans* neuromuscular junctions and is the neurotransmitter used by one-third of neurons in the *C. elegans* nervous system [100]. Cholinergic neurons are involved in many behaviors in *C. elegans*, including locomotion, egg laying, feeding and mating. Sublethal concentrations of the organophosphorous insecticide monocrotophos, still in use in developing countries, caused paralysis, reactive oxygen species production and reduced brood size in *C. elegans* exposed for 4 hours [95]. Chlorpyrifos, a potent acetylcholinesterase inhibitor, also affected brood size after a 72 hour exposure [101]. Dithiocarbamate fungicides like Maneb and Mancozeb, at sublethal concentrations, disrupt dopamine-mediated behaviors [98] and lead to dopamine neurodegeneration [97, 98], which supports epidemiological evidence for a link between carbamate pesticide exposure and Parkinson's disease. Exposure to Mancozeb inhibited locomotion in a dose-dependent manner after a 6 hour exposure [98]. In addition, dopamine-specific behaviors like the transition from swimming to crawling were much more vulnerable to Mancozeb than other behaviors mediated by distinct neuronal subtypes [98]. Dopamine-mediated behaviors were disrupted at the lowest concentrations of Mancozeb, whereas behaviors mediated by other neurons, like egg-laying behavior, mediated by serotonergic neurons, were disrupted at slightly higher concentrations of Mancozeb [98]. Importantly, the behavioral deficits are more sensitive to the compounds and the deficits precede overt neurodegeneration, suggesting that behavioral effects of pesticides serve as the earliest biomarker of pesticide toxicity. *C. elegans* have also been useful for determining gene expression changes due to pesticide exposure [86].

Long-Term Consequences

Why are behavioral biomarkers desirable when evaluating pesticides for toxicity? First, behavioral responses indicate that the organism is responding to a pesticide. Behavioral changes might well be the first indication of impending toxicity to cells and tissues. Second, some behavioral changes might be reversible, suggesting that eventual damage to the underlying cells/tissues/systems that underlie the behavior can be avoided. Third, behavioral effects, particularly in well-studied organisms, can shed light on mechanistic pathways and systems. Finally, behavioral effects provide insight into population and ecological consequences of long-term exposures.

Persistent changes in reproductive behavior, social behavior or predator avoidance behavior could have long-term consequences for an organism's overall fitness. In one study of the anthropogenic pollution effects on fish communication, researchers examined the turbidity of water on coloration and courtship behavior in Lake Victoria cyclids.

Many fish use visual signals to initiate appropriate courtship behavior [102]. The coloration of scales is an important cue for reproductive fitness [102]. In turbid waters, caused by eutrophication and algal blooms or excessive turbulence or increased sediments suspended in the water, these visual communication cues are compromised, which can have an impact on reproduction and possibly overall population characteristics. Over several generations in the turbid water environment, the strong coloration patterns of male cyclids disappeared and the frequency of more mottled, intermediate coloration patterns increased [102]. In another study, endocrine-disrupting properties of a fungicide led to a reduction in the orange coloration spots of male guppies, which in turn reduced the mating success of those males [78]. Male stickleback fish exposed to pesticides showed reduced male nest building and aggressiveness [103], which compromised their reproductive success. Fish grown in waters polluted with mixtures of pesticides and other chemicals exhibited changes in aggressive behavior and predator avoidance escape behavior across generations, indicating heritable genetic or epigenetic mechanisms that could have very long-lasting consequences.

Behavioral changes can be due to sensory system dysfunction, to physiological responses or to developmental effects. These changes can differ depending on when an organism is exposed to the chemicals. Chlorpyrifos exposure, at low doses such as measured in the environment, during development of the nervous system, can profoundly affect behavior and susceptibility to disease later in life. Chlorpyrifos at levels below those required to inhibit the intended target acetylcholinesterase, affect the neuroendocrine regulation of development of sexually dimorphic behaviors in rodents. When exposed pre-natally to chlorpyrifos, adult female mice were less aggressively protective of their own nest, but exhibited enhanced maternal behavior towards the pups, as well as increased anxiety behavior [45]. Prenatally-exposed males exhibited increased aggressive behaviors and increased locomotory activity. These social behavior differences correlated with developmental differences in key neurotransmitter and neuroendocrine brain regions that persisted throughout the lifespan.

Courtship Behavior

Pesticides can have substantial effects on reproductive success in non-target organisms, which could potentially have considerable negative impacts on wildlife. For example, the highly imperiled Buff-breasted sandpiper (*Tryngites subruficollis*) is a migratory shorebird that stops over in agricultural fields in the Rainwater Basin area of Nebraska [104]. During their spring migration stopover, these birds forage, rest and engage in social and courtship behaviors that could be impaired by exposure to pesticides. Migratory sea birds, including the albatross and petrel, have measurable levels of organochlorine pesticides in fatty tissues [105]. Flying ability was affected in homing pigeons exposed experimentally to environmentally relevant amounts of chlorpyrifos and aldicarb [106]. An experimental study of chronic low-dose exposure to the organophosphorous pesticide fenitrothion, affected several key reproductive behaviors in male three-spined stickleback fish [107]. Fenitrothion, which in low concentrations is a potent anti-androgen, caused a reduction in male nest-building and impairments in the zigzag courtship dance, both behaviors regulated by testosterone. Male spiders (*Rabidosa rabida*) exposed to agriculturally relevant concentrations of malathion were unable to perform key courtship behaviors effectively, leading to the males being killed by female spiders prior to copulation [108]. From birds to fish to non-target arthropods, increasing evidence suggests that non-target organisms suffer wide-spread behavioral impairments that could adversely affect population dynamics and success, which could broadly affect numerous ecosystems, even those far away from the site of use of the pesticides.

Maternal Behavior

Low concentrations of many pesticides, including organophosphorous, organochlorine and carbamate compounds, affect diverse hormone-mediated systems, thus acting like endocrine disrupting chemicals. Even if a compound does not directly interact with steroid hormone receptors, many systems regulated or mediated by steroid hormones can be influenced via pesticide effects on signal transduction pathways, effects on gene expression or effects on metabolic pathways. Animals exposed to low, chronic levels of pesticides, far below the threshold for their intended effects, exhibit impairments in reproductive behaviors and other sexually dimorphic behaviors, particularly if animals are exposed to pesticides *in utero* or early in development. Chlorpyrifos exposure neonatally has been shown to inhibit DNA synthesis, neuronal differentiation and synaptogenesis in rats [109]. Long-term changes in sexually dimorphic behaviors measured in adult animals exposed perinatally to low levels of chlorpyrifos [110,40] were associated with changes in the development of neural systems, including nontarget serotonergic systems [110]. Both males and females are affected, often in different ways. For example, adult female mice exposed to chlorpyrifos in utero had reduced anxiety behavior and reduced aggression, while adult males showed increased anxiety behavior and aggression. Pre-natally exposed adult females also showed changes in maternal behaviors towards their own young, including reduced nest defense, reduced anxiety in the presence of a strange male approaching the nest and increased licking and crouching of the litter [45]. Adult female mice exposed to the pesticide 2,4 D spent less time nursing their litters even while increasing licking of pups. As a result, litter growth was impaired by the overall reduced time spent nursing. Other animals show impairments in maternal behavior as a result of pesticide exposure: birds living in pesticide-contaminated areas showed parental neglect [111]. Impairments in maternal care can have long-term effects on the offspring, leading to trans-generational effects that are both physiological via developmental changes in the offspring and behavioral as a result of impairments in early socialization learning [45].

ACUTE VERSUS PERSISTENT EXPOSURES

An important issue that has received little direct experimental attention is the differences in response to an environmental contaminant when exposed acutely as compared to living an entire lifespan in the presence of the contaminant. When organisms or populations of organisms are exposed to chemicals throughout their lifespan, physiological and genetic adaptations are possible that would reflect different behavioral effects than in organisms newly exposed to a compound. For example, we found that adult nematodes exposed for 6 hours to

Mancozeb exhibited profound locomotory impairment, becoming paralyzed or barely moving and curling up tightly. In addition, their bodies appeared swollen. However, with a 24 hour exposure, those surviving worms exhibited normal locomotory behavior [98] and their bodies were not swollen or misshapen. Offspring exposed throughout development behaved normally as adults (Raley-Susman et al, unpublished observations). A similar result was seen in *Daphnia magna* exposure for two generations to an environmentally relevant mixture of pesticides [112]. Many studies have demonstrated that exposure during development to low levels of pesticides that can act as endocrine disrupters leads to life long changes in neuroendocrine function, reproductive and sexually dimorphic behaviors [36]. There is also increased risk of developing cancer in childhood or adulthood [113].

Several studies have reported persistent behavioral changes, changes that appear to be independent of targeted effects on acetylcholinesterase inhibition, in rodents exposed to low doses of organophosphate pesticides [114] or carbamate fungicides [115]. Repeated exposures to doses of chlorpyrifos that are below threshold for acetylcholinesterase inhibition caused attention deficits and increased impulsive behavior in rats [116]. Further, long-term exposure of zebra fish, *Danio rerio*, to sublethal concentrations of parathion in the water, led to an increase in general motor activity and food consumption [117]. The behavioral changes occurred in the absence of overt neurological impairment [118] in one study of chronic exposure of mice to rotenone.

MIXTURES

Most organisms, including humans, are exposed to a complex mixture of many different pesticides at once, at varying concentrations and for varying lengths of time. Most freshwater sources contain hundreds of measurable human-produced chemicals. Of the sixteen contaminants present at the highest concentrations in Lake Michigan, eleven of them are pesticides, including diazinon, chlorpyrifos, endosulfan, melathion, atrazine, permethrins, dichlorvos, manganese, zinc, imidacloprid and naphthalenes [119]. Given that the Great Lakes provide 20% of the Earth's freshwater supply, the effects of these contaminants on the organisms that depend on this water are of critical importance to understand. Depending on the concentrations organisms are exposed to, the mechanisms of action can vary from effects on neuromuscular function to effects on other neurotransmitter systems, hormone systems, cell growth and metabolism. In addition, mixtures could act synergistically or additively. For example, a combination of three common fungicides used routinely in the wine industry acted synergistically at low concentrations to cause greater neurotoxicity than the compounds exerted individually [120]. In another study, long-term exposure to low dose concentrations of two different pesticides, chlorpyrifos and the pyrethroid deltamethrin, caused multiple effects in rat brain tissue. Some effects were additive and some were synergistic [121]. Similarly, azole fungicides have been shown to enhance the toxicity of pyrethroid insecticides to aquatic invertebrates both under laboratory and field conditions [122]. Further, atrazine has been shown to enhance the toxicity of organophosphorus pesticides [123]. Much more work is needed to understand the effects of the complex mixtures of different pesticides and other human contaminants on organismal behavior, fitness and health. It is abundantly clear, however, that the extensive use of pesticides is altering the behavior and possible fitness of many organisms across the globe and in all ecosystems. Because of the persistence of these compounds and the long-term consequences of exposure, some of which are trans-generational, we need to acknowledge that life in a world of pesticides is the new normal.

SUMMARY AND CONCLUSIONS

Behavioral neurotoxicological and behavioral ecotoxicological approaches need to take prominence in evaluating the short and long-term consequences to wildlife and humans of pesticide use. Behavior can reveal much about the systems and processes affected by pesticides, as well as the mechanisms by which pesticides exert those effects. Behavioral change is often the earliest sign of harmful effects of pesticides, and even low doses of pesticides or pesticide mixtures can lead to long-term behavioral change, particularly when exposures occur during developmental sensitive periods. Because we are living in a world with measureable mixtures of pesticides and other human-produced chemicals, it is essential that more work be done to understand how animals, including humans, are living and behaving in this new surrounding milieu.

ACKNOWLEDGEMENTS

Thanks to Ms. Melanie Susman for the artwork of Figure 1.

REFERENCES

1. Hazell, P.B.R. 2009 The Asian Green Revolution. In: International Food Policy Research Institute Discussion Paper 00911. P1-40.
2. Hirsch, N.P. Neuromuscular junction in health and disease. British Journal of Anaesthesia 2007; 99(1) 132-8.
3. Lindstrom, J. Nicotinic acetylcholine receptors of muscles and nerves: comparison of their structures, functional roles and vulnerability to pathology. Annals of the New York Academy of Science 2003; 998: 41-52.
4. Massoulie, J., Millard, C.B. Cholinesterase and the basal lamina at vertebrate neuromuscular junctions. Current Opinions in Pharmacology 2009; 9: 316-325.
5. Brandan, E, Maldonado, M, Garrido, J, Inestrosa, NC. Anchorage of collagen-tailed acetylcholinesterase to the extracellular matrix is mediated by heparin sulfate proteoglycans. Journal of Cell Biology 1985; 101: 985-992.
6. Massoulie, J. The origin of the molecular diversity and functional anchoring of cholinesterases. NeuroSignals 2002; 11: 130-143.
7. Belanger, JH. Contrasting tactics in motor control by vertebrates and arthropods. Integrative and Comparative Biology 2005; 45:672-678.
8. Rathmayer, W, Erxleben, C. Identified muscle fibers in a crab. I. Characteristics of excitatory and inhibitory neuromuscular transmission. Journal of Comparative Physiology 1983; 152:411-420.
9. Cook, C. Earth Island Journal, 2005. http://www.thirdworldtraveler.com/Environment/Spraying-America.html (accessed 20 August 2013).

10. Casida, JE, Nomura, DK, Vose, SC, Fujioka, K. Organophosphate-sensitive lipases modulate brain lysophospholipids, ether lipids and endocannabinoids. Chemico-Biological Interactions 2008; 175: 355-364.
11. Pope, CN. Organophosphorus pesticides: Do they all have the same mechanism of toxicity? Journal of Toxicological and Environmental Health B Critical Reviews 1999; 2: 161-181.
12. Haynes, KF. Sublethal effects of neurotoxic insecticides on insect behavior. Annual Review of Entomology 1988; 33:149-168.
13. Source.epa.gov/pesticides/ecosystem/ecorisk.htm
14. Dominguez, I, Agra, AR, Monaghan, K, Soares, AMVM, Noguiera, AJA. Cholinesterase and glutathione-s-transferease activities in freshwater invertebrates as biomarkers to assess pesticide contamination. Environmental Toxicology and Chemistry 2010; 29(1) 5-18.
15. Ruark, CD, Hack, CE, Robinson, PJ, Gearhart, JM. Quantitative structure-activity relationships for organophosphates binding to trypsin and chymotrypsis. Journal ofToxicology and Environmental Health A: Current Issues 2011; 74(1) 1-23.
16. Woodruff, TJ, Zota, AR, Schwartz, JM. Environmental chemicals in pregnant women in the United States: NHANES 2003-4. Environmental Health Perspectives 2011; 119:878-885.
17. Hertz-Picciotto, I, Delwiche, L. The rise in autism and the role of age at diagnosis. Epidemiology 2009; 20(1) 84-90.
18. Roberts, EM, English, PB, Grether, JK, Windham, GC, Somberg, L, Wolff, C. Maternal residence near agricultural pesticide applications and autism spectrum disorders among children in the California Central Valley. Environmental Health Perspectives. 2007; 115: 1482-1489.
19. Eskenazi, B, Marks, AR, Bradman, A, Harley, K, Barr, DB, Johnson, C, et al. Organophosphate pesticide exposure and neurodevelopment in young Mexican-American children. Environmental Health Perspectives 2007; 115: 792-798.
20. Shelton, JF, Hertz-Picciotto, I, Pessah, IN. Tipping the balance of autism risk: Potential mechanisms linking pesticides and autism. Environmental Health Perspectives 2012; 120(7) 944-951.

21. Levin, ED, Addy, N, Nakajima, A, Christopher, NC, Seidler, FJ, Slotkin, TA. Persistent behavioral consequences of neonatal chlorpyrifos exposure in rats. Developmental Brain Research 2001; 130(1) 83-89.
22. Aluigi, MG, Angelini, C, Falugi, C, Fossa, R, Genever, P, Gallus, L. Interaction between organophosphate compounds and cholinergic functions during development. Chemical and Biological Interactions 2005; 157-8: 305-316.
23. Yang, D, Lauridsen, H, Buels, K, Chi, LH, La Du, J, Bruun, DA. Chlorpyrifos-oxon disrupts zebra fish axonal growth and motor behavior. Toxicological Sciences 2011; 121(1) 146-159.
24. Read, DJ, Li,Y, Chao, MV, Cavanagh, JB, Glynn, P. Organophosphates induce distal axonal damage, but not brain oedema, by inactivating neuropathy target esterase. Toxicology and Applied Pharmacology 2010; 245: 108-115.
25. Crow, JA, Bittles, V, Herring, KL, Borazjani, A, Potter, PM, Ross, MK. Inhibition of recombinant human carboxylesterase 1 and 2 and monoacylglycerol lipase by chlorpyrifos oxon, paraoxon and methyl paraoxon. Toxicology and Applied Pharmacology 2012; 258: 145-50.
26. Galloway, T, Handy, R. Immunotoxicity of organophosphorous pesticides. Ecotoxicology 2003; 12:345-363.
27. Stepanovic, RM, Jokanovic, M, Maksimovic, M. Toxicological importance of lymphocyte neuropathy target esterase. Archives in Pharmacology. 1998; 358(2) 5413-5418.
28. Casale, GP, Cohen, SD, Dicapua, RA. The effects of organophosphate-induced stimulation on the antibody response to sheep erythrocytes in inbred mice. Toxicology and Applied Pharmacology. 1983; 68: 198-205.
29. Rodgers, KE, Ellefson, DD. Modulation of respiratory burst activity and mitogen response of human peripheral blood mononuclear cells and murine splenocytes and peritoneal cells by malathion. Fundamentals of Applied Toxicology 1990; 14: 309-17.
30. Rodgers, KE, Ellefson, DD. Mechanism of modulation of murine peritoneal cell function and mast cell degranulation by low doses of malathion. Agents Actions 1992; 35:57.

31. Eason, CT, Svendsen, C, O'Halloran, K, Weeks, JM. An assessment of the lysosomal neutral red retention test and immune function assay in earthworms (*Eisinia andrei*) following exposure to chlorpyrifos, benzo-a-pyrene and contaminated soil. Pedobiologia 1999; 43:641-5.

32. Booth, LH, O'Halloran, K. A comparison of biomarker responses in the earthworm *Aporrectodea caliginosa* to the organophorphorous insecticides diazinon and chlorpyrifos. Environmental Toxicological Chemistry. 2001; 20: 2494-502.

33. Beaman, JR, Finch, R, Gardner, H, Hoffmann, F, Rosencrance, A, Zelikoff, JT. Mammalian immunoassays for predicting the toxicity of malathion in a laboratory fish model. Journal of Toxicology and Environmental Health A 1999; 56: 523-42.

34. Khalaf-Allah, SS. Effect of water pollution on some haematological, biochemical and immunological parameters in *Tilapia nilotica* fish. Deutche Tierarztliche Wochenschrift 1999; 106: 67-71.

35. Crumpton, TL, Seidler, FJ, Slotkin, TA. Developmental neurotoxicity of chlorpyrifos: In vivo and in vitro effects on nuclear transcription factors involved in cell replication and differentiation. Brain Research 2000; 857: 87-98.

36. Zala, SM, Penn, DJ. Abnormal behaviours induced by chemical pollution: a review of the evidence and new challenges. Animal Behaviour 2004; 68: 649-664.

37. Levin, ED, Addy, N, Baruah, A, Elias, A, Christopher, N, Seidler, FJ, Slotkin, TA. Prenatal chlorpyrifos exposure in rats causes persistent behavioral alterations. Neurotoxicology and Teratology. 2002; 24:733-741.

38. Slotkin, TA, Cousins, ML, Tate, CA, Seidler, FJ. Persistent cholinergic presynaptic deficits after neonatal chlorpyrifos exposure. Brain Research 2001; 902: 229-243.

39. Slotkin, TA, Tate, CA, Cousins, MM, Seidler, FJ. Functional alterations in CNS catecholamine systems in adolescence and adulthood after neonatal chlorpyrifos exposure. Developmental Brain Research. 2002; 287: 163-173.

40. Icenogle, LM, Christopher, NC, Blackwelder, WP, Caldwell, DP, Qiao, D, Seidler, FJ, Slotkin, TA, Levin, ED. Behavioral alterations in adolescent and adult rats caused by a brief subtoxic exposure to

chlorpyrifos during neurulation. Neurotoxicology and Teratology. 2004; 26: 95-101.

41. Frye, CA, Bo, E, Calamandrei, G, Calza, L, Dessi-Fulgheri, F, Fernandez, M, Fusani, L, Kah, O, Kajta, M, Le Page, Y, Patisaul, HB, Venerosi, A, Wojtowicz, AK, Panzica, GC. Endocrine disrupters: A review of some sources, effects and mechanisms of actions on behavior and neuroendocrine systems. Journal of Neuroendocrinology. 2011; 24:144-159.

42. Scholz, NL, Truelove, NK, French, BL, Berejikian, BA, Quinn, TP, Sacillas, E and Collier, TK. Diazinon disrupts antipredator and homing behaviors in Chinook salmon (*Oncorhynchus tsawytscha*). Canadian Journal of Fisheries and Aquatic Sciences. 2000; 57: 1911-1918.

43. Tait, S, Ricceri, L, Venerosi, A, Maranghi, F, Mantovani, A, Calamandrei, G. Long-term effects on hypothalamic neuropeptides after developmental exposure to chlorpyrifos in mice. Environmental Health Perspectives. 2009; 117: 112-116.

44. Ricceri, L, Markina, N, Valanzano, A, Fortuna, S, Cometa, MF, Maneguz, A, Calamandrei, G. Developmental exposure to chlorpyrifos alters reactivity to environmental and social cues in adolescent mice. Toxicology and Applied Pharmacology. 2003; 191: 189-201.

45. Venerosi, A, Ricceri, L, Tait, S, Calamandrei, G. Sex-dimorphic behaviors as markers of neuroendocrine disruption by environmental chemicals: The case of chlorpyrifos. Neurotoxicology. 2012; 33: 1420-1426.

46. Norstrom,RJ, Belikov, SE, Born, EW, Garner, GW, Malone, B, Olpinski, S, Ramsay, MA, Schliebe, S, Stirling, I, Stishov, MS, Taylor MK, Wiig, O. Chlorinated hydrocarbon contaminants in polar bears from eastern Russia, North America, Greenland, and Svalbard: Biomonitoring of Arctic pollution. Archives of Environmental Contaminants and Toxicology. 1998; 35:354-367.

47. Silva, MH, Gammon, D. An assessment of the developmental, reproductive, and neurotoxicity of endosulfan. Birth Defects Research (Part B). 2009; 86: 1-28.

48. Davies, TGE, Field, LM, Usherwood, PNR, Williamson, MS. DDT, Pyrethrins, Pyrethroids and insect sodium channels. IUBMB Life 2007; 59(3) 151-162.

49. Sunderam, RIM, Thomson, GB, Cheng, DMH. Toxicity of endosulfan to native and introduced fish in Australia. Environmental and Toxicological Chemistry 1992; 11(10) 1469-1476.

50. Jia, H, Liu, L, Sun, Y, Cai, D, Hu, J, Ren, N, Li, Y. Endosulfan in the Chinese environment: monitoring and modeling. Frontiers in Environmental Science Engineering 2012; 6(1) 32-44.

51. Park, D, Hempleman, SC, Propper, CR. Endosulfan exposure disrupts pheromonal systems in the red-spotted newt: a mechanism for subtle effects of environmental chemicals. Environmental Health Perspectives 2001; 109: 669-673.

52. Vom Saal, FS, Nagel, SC, Palanza, P, Boechler, M, Parmigiani, S, Welshons, WV. Estrogenic pesticides: binding relative to estradiol in MCF-7 cells and effects of exposure during fetal life on subsequent territorial behavior in male mice. Toxicological Letters. 1995; 77: 343-350.

53. Palanza, P, Morellini, F, Parmigiani, S, vom Saal, FS. Ethological methods to study the effects of maternal exposure to estrogenic endocrine disrupters: a study with methoxychlor. Neurotoxicology and Teratology 2002; 24:55-69.

54. Berrill, M, Coulson, D, McGillivray, L, Pauli, B. Toxicity of endosulfan to aquatic stages of anuran amphibians. Environmental and Toxicological Chemistry. 1998; 17: 1738-1744.

55. Morais, S, Dias, E, Pereira, M. Carbamates: Human exposure and health effects. In: The Impact of Pesticides. AcademyPublish.org. 2012; p. 21-38.

56. Weichenthal, S, Moase, C, Chan, P. A review of pesticide exposure and cancer incidence in the Agricultural Health Study cohort. Environmental Health Perspectives. 2010; 118: 1117-1125.

57. Frazier, LM. Reproductive disorders associated with pesticide exposure. Journal of Agromedicine 2007; 12(1) 27-37.

58. Meeker, JD, Singh, NP, Ryan, L, Duty, SM, Barr, DB, Herrick, RF, Bennett, DH, Hauser, R. Urinary levels of insecticide metabolites and DNA damage in human sperm. Human Reproduction. 2004; 19(11) 2573-2580.

59. Wang, A, Costello, S, Cockburn, M, Zhang, X, Bronstein, J, Ritz, B. Parkinson's disease risk from ambient exposure to pesticides. European Journal of Epidemiology. 2011; 26: 547-55.

60. Easton, A, Guven, K, dePomerai, DI. Toxicity of the dithiocarbamate fungicide Mancozeb to the nontarget soil nematode, *Caenorhabditis elegans*. Journal of Biochemical and Molecular Toxicology. 2001; 15(1) 15-25.

61. Benedetto,A, Au,C, Avila,DS, Milatovic,D, Aschner,M. Extracellular dopamine potentiates Mn-induced oxidative stress, lifespan reduction, and dopaminergic neurodegeneration in a BLI-3-dependent manner in *Caenorhabditis elegans*. PLOSGenetics 2010; 6:1-17.

62. Freire, C, Koifman, S. Pesticide exposure and Parkinson's disease: Epidemiological evidence of association. NeuroToxicology. 2012; 33: 947-971.

63. Steinrucken, HC, Amrhein, N. The herbicide glyphosate is a potent inhibitor of 5-enol-pyruvylshikimic acid-3-phosphate synthase. Biochemical and Biophysical Research Communications 1980; 94(4) 1207-12.

64. Giesy, JP, Dobson, S, Solomon, KR. Ecotoxicological risk assessment of Roundup herbicide. IN: Reviews of Environmental Contamination and Toxicology 2000; 167: 35-120.

65. Walsh, LP, McCormick, C, Martin, C, Stocco, DM. RoundUp inhibits steroidogenesis by disrupting a steroidogenic acute regulatory (StAR) protein expression. Environmental Health Perspectives 2000; 108(8) 769-76.

66. Romano, RM, Romano, M, Bernardi, MM, Furtado, PV, Oliveira, CA. Prepubertal exposure to commercial formulations of the herbicide glyphosate alters testosterone levels and testicular morphology. Archives of Toxicology 2009; 84(4) 309-17.

67. Lammer, EJ, Chen, DT, Hoar, RM, Agnish, ND, Benke, PJ, Braun, JT, Curry, CJ, Femhoff, PM. Retinoic acid embryopathy. New England Journal of Medicine. 1985; 313(4) 837-41.

68. Paganelli, A, Gnazzo, V, Acosta, H, Lopez, SL, Carrasco, AE. Glyphosate-based herbicides produce teratogenic effects on vertebrates by impairing retinoic acid signaling. Chemical Research in Toxicology. 2010; 23(10) 1586-9.

69. Richard, S, Moslemi, S, Sipahutar, H, Benachour, W, Seralini, G. Differential effects of glyphosate and RoundUp on human placental cells and aromatase. Environmental Health Perspectives. 2005; 113(6) 716-20.

70. Kollmeyer, WD, Flattum, RF, Foster, JP, Powell, JE, Schroeder, ME, Soloway, SB. Discovery of the nitromethylene heterocycle insecticides. In: Yamamoto, I, Casida,J (ed) Nicotinoid insecticides and the nicotinic acetylcholine receptor. Tokyo. Springer-Verlag. P71-89.

71. Tomizawa, M, Latili, B, Casida, JE. Structure and function of insect nicotinic acetylcholine receptors studied with nicotinic insecticide affinity probes. In: Yamamoto, I, Casida,J (ed) Nicotinoid Insecticides and the Nicotinic Acetylcholine Receptor. Tokyo. Springer-Verlag. p. 271-92.

72. Whitehorn, PR, O'Connor, S, Wackers, FL, Goulson, D. Neonicotinoid pesticide reduces bumble bee colony growth and queen production. Science 2012; 336: 351-2.

73. Williamson, SM, Wright, GA. Exposure to multiple cholinergic pesticides impairs olfactory learning and memory in honeybees. Journal of Experimental Biology. 2013; 216: 1799-1807.

74. Nyman, A, Hintermeister, A, Schirmer, K, Aschauer, R. The insecticide imidacloprid causes mortality of the freshwater amphipod *Gammarus pulex* by interfering with feeding behavior. PLOS One 8: e62472. Doi: 10.1371/journal.pone.0062472.

75. Capowiez, Y, Berard, A. Assessment of the effects of imidacloprid on the behavior of two earthworm species (*Aporrectodea nocturna* and*Allolobophora icterica*) using 2D terraria. Ecotoxicological and Environmental Safety 2006; 64: 198-206.

76. Tisler, T, Jemec, A, Mozetic,B, Trebse, P. Hazard identification of imidacloprid to aquatic environment. Chemosphere 2008; 76: 907-14.

77. Broley, C. The plight of the American bald eagle. Audobon Magazine 1958; 60: 162-3, 171.

78. Baatrup, E, Junge, M, Gunier, RB, Harnly, ME, Reynolds, P, Hertz, A, Von Behren, J. Antiandrogenic pesticides disrupt sexual characteristics in the adult male guppy *Poecilia reticulates*. Environmental Health Perspectives. 2001; 109: 1063-1070.

79. Tierney, KB, Sampson, JL, Ross, PS, Sekela, MA, Kennedy, CJ. Salmon olfaction is impaired by an environmentally realistic pesticide mixture. Environmental Science Technology. 2008; 42: 4996-5001.
80. Bell, AM. Effects of an endocrine disrupter on courtship and aggressive behavior of male three-spined stickleback, *Gasterosteus aculeatus*. Animal Behaviour. 2001; 62: 775-780.
81. Porter, WP, Jaeger, JW, Carlson, IH. Endocrine, immune, and behavioural effects of aldicarb (carbamate), atrazine (triazine) and nitrate (fertilizer) mixtures at groundwater concentrations. Toxicology and Individual Health 1999; 15: 133-150.
82. Relyea, RA. New effects of Roundup on amphibians: Predators reduce herbicide mortality; herbicides induce antipredator morphology. Ecological Applications 2012; 22(2) 634-647.
83. Holmes, SB, Boag, PT. Effects of the organophosphorus pesticide fenitrothion on behavior and reproduction in zebra finches. Environmental Research 1990; 53: 62-75.
84. Flynn, KM, Delclos, KB, Newbold, RR, Ferguson, SA. Behavioural responses of rats exposed to long-term dietary vinclozolin. Journal of Agricultural Food Chemistry. 2001; 49: 1658-1665.
85. Little, EE, Finger, SE. Swimming behavior as an indicator of sublethal toxicity in fish. Environmental and Toxicological Chemistry. 1990; 9:13-19.
86. Rand, MD. The growing potential of *Drosophila* in neurotoxicology. Neurotoxicology and Teratology. 2010; 32: 74-83.
87. Leung, MCK, Williams, PL, Benedetto, A, Au, C, Helmcke, KJ, Aschner, M, Meyer, JN. *Caenorhabditis elegans*: An emerging model in biomedical and environmental toxicology. Toxicological Sciences 2008; 106(1) 5-28.
88. Anderson, GL, Cole, RD, Williams, PL. Assessing behavioral toxicity with *Caenorhabditis elegans*. Environmental and Toxicological Chemistry. 2004; 23:1235-1240.
89. Boyd, WA, Cole, RD, Anderson, GL, Williams, PL. The effects of metals and food availability on the behavior of *Caenorhabditis elegans*. Environmental and Toxicological Chemistry. 2003; 22: 3049-3055.

90. Liou, HH, Tsai, MC, Chen, CJ, Jeng, JS, Chang, YC, Chen, SY, Chen, RC. Environmental risk factors and Parkinson's disease: A case-control study in Taiwan. Neurology 1997; 48: 1583-1588.
91. Kim, Y, Sun, H. Functional genomic approach to identify novel genes involved in the regulation of oxidative stress resistance and animal lifespan. Aging 2007; 6:489-503.
92. Nonet, ML, Saifee, O, Zhao, H, Rand, JB, Wei, L. Synaptic transmission deficits in *Caenorhabditis elegans* synaptobrevin mutants. Journal of Neuroscience. 1998; 18(1) 70-80.
93. Wu, Z, Smith, JV, Paramashivarn, V, Butko, P, Khan, I, Cypser, JR, Luo, Y. Ginko biloba extract EGb 761 increases and extends life span of*Caenorhabditis elegans*. Cell and Molecular Biology 2002; 48(6) 725-731.
94. Hartman, P, Childress, E, Beyer, T. Nematode development is inhibited by methyl viologen and high oxygen concentration at a rate inversely proportional to lifespan. J. Gerontol A Biol Sci Med Sci 1995; 50: B322-B326.
95. Leelaja, BC, Rajini, PS. Biochemical and physiological responses in *Caenorhabditis elegans* exposed to sublethal concentrations of the organophosphorus insecticide, monocrotophos. Ecotoxicology and Environmental Safety 2013; 94: 8-13.
96. Roh, JY, Choi, J. Ecotoxicological evaluation of chlorpyrifos exposure on the nematode *Caenorhabditis elegans*. Ecotoxicology and Environmental Safety 2008; 71: 483-489.
97. Negga, R, Stuart, JA, Machen, ML, Salva, J, Sizek, AJ, Rishardson, SJ. Exposure to glyphosate-and/or Mn/Zn-ethylene-bis-dithiocarbamate-containing pesticides leads to degeneration of g-aminobutyric acid and dopamine neurons in *Caenorhabditis elegans*. Neurotoxicology Research. 2012; 21: 281-290.
98. Brody, AH, Chou, E, Gray, JM, Pokrywka, NJ, Raley-Susman, KM. Mancozeb-induced behavioral deficits precede structural neuronal degeneration. NeuroToxicology. 2013; 34:74-81.
99. Williams, PL, Dusenbery, DB. A promising indicator of neurobehavioral toxicity using the nematode *Caenorhabditis elegans* and computer tracking. Toxicology and Individual Health. 1990; 6: 425-440.

100. Rand, JB. Acetylcholine. Wormbook 2007 (http://www.wormbook.org/chapters/www_acetylcholine/acetylcholine.html)

101. Ruan, QL, Ju, JJ, Li, YH, Liu, R, Pu, YP, Yin, LH, Wang, DY. Evaluation of pesticide toxicities with differing mechanisms using *Caenorhabditis elegans*. Journal of Toxicology and Environmental Health Part A 2009; 72: 746-751.

102. Van der Sluijs, I, Gray, SM, Amorim, MCP, Barber, I, Candolin, U, Hendry, AP, Krahe, R, Maan, ME, Utne-Palm, AC, Wagner, HJ, Wong, BBM. Communication in troubled waters: responses of fish communication systems to changing environments. Evolutionary Ecology 2011; 25: 623-640.

103. Breckels, RD, Neff, BD. Pollution-induced behavioural effects in the brown bullhead (*Ameiurus nebulosus*). Ecotoxicology 2010; 19:1337-1346.

104. McCarty, JP, Jorgensen, JG, Wolfenbarger, LL. Behavior of buff-breasted sandpipers (Tryngites subruficollis) during migratory stopover in agricultural fields. PLOS One 2009; 4(11) e8000. doi:10.1371/journal.pone.0008000

105. Colabuono, FI, Taniguchi, S, Montone, RC. Organochlorine contaminants in albatrosses and petrels during migration in South Atlantic Ocean. Chemosphere 2012; 86: 701-8.

106. Moye, JK, Pritsos, CA Effects of chlorpyrifos and aldicarb on flight activity and related cholinesterase inhibition in homing pigeons, *Columba livia*: Potential for migration effects. Bulletin of Environmental Contamination and Toxicology 2010; 84:677-681.

107. Sebire, M, Scott, AP, Tyler, CR, Cresswell, J, Hodgson, DJ, Morris, S, Sanders, MB, Stebbing PD, Katsiadaki, I. The organophosphorus pesticide, fenitrothion, acts as an anti-androgen and alters reproductive behavior of the male three-spined stickleback, *Gasterosteus aculeatus*. Ecotoxicology 2009; 18: 122-133.

108. Tietzen, WF. Pesticides affect the mating behavior of *Rabidosa rabida* (Araneae, Lycosidae). Journal of Arachnology 2006; 34: 285-288.

109. Crumpton, TL, Seidler, FJ, Slotkin, TA. Developmental neurotoxicity of chlorpyrifos in vivo and in vitro: effects on

nuclear transcription factors involved in cell replication and differentiation. Brain Research 2000; 857: 87-98.

110. Aldridge, JE, Livin, Ed, Seidler, FJ, Slotkin TA. Developmental exposure of rats to chlorpyrifos leads to behavioral alterations in adulthood, involving serotonergic mechanisms and resembling animal models of depression. Environmental Health Perspectives 2005; 113: 527-531.
111. Clotfelter, ED, Bell, AM, Levering, KR. The role of animal behavior in the study of endocrine-disrupting chemicals. Animal Behaviour 2004; 68: 665-676.
112. Brausch, JM, Salice, CJ. Effects of an environmentally realistic pesticide mixture on Daphnia magna exposed for two generations. Archives of Environmental Contamination and Toxicology 2011; 61: 272-279.
113. Alavanja, MCR. Pesticides use and exposure extensive worldwide. Reviews in Environmental Health 2009; 24(4) 303-309.
114. Timofeeva, OA, Sanders, D, Seemann, K, Yang, L, Hermanson, D, Regenbogen, S, Agoos, S, Kallepalli, A, Rastogi, A, Braddy, D, Wells, C, Perraut, C, Seidler, FJ, Slotkin, TA, Levin, ED. Persistent behavioral alterations in rats neonatally exposed to low doses of the organophosphate pesticide, parathion. Brain Research Bulletin 2008; 77: 404-411.
115. Jia, Z, Misra, HP. Developmental exposure to pesticides zineb and/or endosulfan renders the nigrostriatal dopamine system more susceptible to these environmental chemicals later in life. NeuroToxicology 2007; 28: 727-735.
116. Middlemore-Risher, ML, Buccafuscu, JJ, Terry, AV. Repeated exposure to low level chlorpyrifos results in impairments in sustained attention and increased impulsivity in rats. Neurotoxicology and Teratology. 2010; 32: 415-424.
117. Roe, X, Keijzers, R, van Gestel, CAM. Acetylcholinesterase inhibition and increased food consumption rate in the zebra fish, *Danio rerio*, after chronic exposure to parathion. Aquatic Toxicology. 2003; 64: 457-460.
118. Richter, F, Hamann, M, Richter, A. Chronic rotenone treatment induces behavioral effects but no pathological signs of Parkinsonism in mice. Journal of Neuroscience Research 2007;

85: 681-691.

119. Illinois Environmental Protection Agency. Lake Michigan Monitoring Program. http://www.epa.state.il.us/water/surface-water/lake-michigan-mon.html (accessed 23 August 2013).

120. Coleman, MD, O'Neil, JD, Woehrling, EK, Ndunge, OBA, Hill, EJ, Menache, A, Reiss, CJ. A preliminary investigation into the impact of a pesticide combination on human neuronal and glial cell lines in vitro. PLOS One 2012; 7(8) c42768. Doi:10.1371/journal.pone.0042768.

121. Tuzmen, MN, Candan, N, Kaya, E. The evaluation of altered antioxidative defense mechanism and acetylcholinesterase activity in rat brain exposed to chlorpyrifos, deltamethrin, and their combination. Toxicology Mechanisms and Methods 2007; 17: 535-40.

122. Bjergager, MBA, Hanson,ML, Solomon, KR, Cedergreen, N. Synergy between prochloraz and esfenvalerate in Daphnia magna from acute and subchronic exposures in the laboratory and microcosms. Aquatic Toxicology. 2012; 110-111: 17-24.

123. Anderson, TD, Lydy, MJ. Increased toxicity to invertebrates associated with a mixture of atrazine and organophosphate insecticides. Environmental Toxicology and Chemistry. 2002; 21: 1507-1514.

Chapter 6

Distribution, Mechanisms, Impact and Management of Insecticide Resistance in Malaria Vectors: A Pragmatic Review

Vincent Corbel[1, 2] and Raphael N'Guessan[3, 4]

[1]Institut de Recherche pour le Développement, Maladies Infectieuses et Vecteurs, Ecologie, Génétique, Evolution et Contrôle (IRD 224-CNRS 5290 UM1-UM2), Benin

[2]Department of Entomology, Kasetsart University, Bangkok, Thailand

[3]London School of Hygiene & Tropical Medicine, London, UK

[4]Centre de Recherches Entomologiques de Cotonou, Cotonou, Benin

INTRODUCTION

Malaria is still a major burden causing the death of nearly 655,000 people each year, mostly in children under the age of five, and affecting those living in the poorest countries [1]. Currently, the major obstacles to malaria control and elimination are the absence of a protective vaccine, the spread of parasite resistance to anti-malarial drugs and the mosquito resistance to insecticides [2]. Controlling mosquito vectors is fundamental to reduce mosquito-borne diseases by targeting vectorial capacity and hence the transmission. Vector control through the use of chemicals for mosquito bed nets and indoor residual spraying is still the cornerstone of malaria prevention [1]. Unfortunately, the extensive use of insecticides since the 1950s has led to the development of strong resistance worldwide hence representing a major public health problem where insecticidal vector control is implemented. Here, we propose to review the current level, distribution and mechanisms of insecticide résistance in malaria vectors and address their impact on the efficacy of vector control interventions. Strategies to prevent and/or delay the spread of insecticide resistance in natural mosquito populations are also discussed.

DEFINITION OF RESISTANCE

According to the *World Health Organization* (WHO), resistance is defined as the ability of an insect to withstand the effects of an insecticide by becoming resistant to its toxic effects by means of natural selection and mutations [3]. This definition differs from that provided by the Insecticide Resistance Action committee (IRAC) (www.irac-online.org) that gathers independent scientists and experts belonging to Agrochemical Companies who define operational (field) resistance as a heritable change in the sensitivity of a pest population that is reflected in the repeated failure of a product to achieve the expected level of control when used according to the label recommendation for that pest species. The IRAC definition, although pragmatic, is less "sensitive" with the scope to implement early *Insecticide Resistance Management* (IRM) strategies in the field. In both cases however, appropriate tools (biological, biochemical and/or molecular) are needed to identify the mechanisms involved and to conduct surveillance at individual and/or population levels [4].

Resistance has been observed in more than 500 insect species worldwide among which more than 50*Anopheles* species (Diptera: Culicidae) are responsible for the transmission of malaria parasites to humans [5]. Resistance is a heritable character that relies on a genetic basis. Resistance results from the selection of a genetic modification in one or several genes occurring by migration and/or mutation. For example, when a mosquito population is exposed to an insecticide A, the individuals having resistant genes to this insecticide A survive and reproduce until the resistant allele becomes almost fixed. The use of insecticides for agricultural purposes and more recently for public health has played pivotal step in the selection of resistance in malaria vectors [6]. Resistance can involve several physiological and/or behavioural changes. Changes in the insecticide target site that reduce its binding to insecticides (known as target-site resistance) is the best understood type of resistance mechanism and molecular diagnostics to detect this resistance mechanism are now integrated into insecticide resistance monitoring strategies in malaria control programmes [7, 8]. Enhanced insecticide metabolism that lowers the amount of insecticide reaching the target site (known as metabolic resistance) is more complex but recent advances have identified key enzymes responsible for insecticide detoxification, paving the way for the development of molecular markers for this type of resistance mechanism [9, 10]. Other physiological changes (e.g. reduce penetration through cuticular resistance) and/or behavioural changes in the mosquito population were identified but their impact on the efficacy of insecticides is still poorly understood.

It is commonly accepted that the enhanced metabolism and target site modifications are responsible for high level of insecticide resistance in malaria vectors. To date, malaria vectors have developed resistance to the main chemical classes used in public health (i.e. pyrethroids, DDT, carbamates and organophosphates) (table 1) and the occurrence of cross-resistance[1] - and multiple resistance[2] -represent a serious threat to achieving the Millennium Development Goals for malaria control (i.e 75% reduction of global malaria cases by 2015). Surveillance and routine monitoring campaigns to assess the level and type of resistance are essential to help Malaria Control Programme (MCPs) to design more effective and sustainable malaria vector control strategies at an operational scale [4].

HISTORY OF RESISTANCE TO PUBLIC HEALTH INSECTICIDES

Since the humans used chemicals for crop protection and/or the prevention of vector borne diseases, cases of resistances have been reported [11, 12]. Insecticides used for malaria control have included organochlorine, organophosphorus, carbamate, and pyrethroid insecticides, with the latter now taking increasing market share for both indoor residual spraying and Long Lasting Insecticidal mosquito Nets (LLINs) programmes [13]. Resistance has naturally tended to follow the use and switches of these insecticides [5].

Table 1: Mechanisms of insect resistance to the main insecticide families of public health interest

Insecticides	Molecular target	Resistance mechanisms		References
		target site	enzymatic mechanisms	
Pyrethroids, type I	sodium channel	Kdr and super Kdr mutations	monooxygenases + esterases	[10,61,62]
Pyrethroids, Type II	sodium channel	Kdr mutations	monooxygenases + esterases	[10,61,62]
Organochlorates	sodium channel	Kdr mutations	GS-tranferases + monooxygenases	[10,61,62]
N-alkyl-amides	sodium channel	no resistance reported against insect of public health importance		
Organophosphates	acetylcholinesterase	$Ace1^R$ mutation	esterases + GS-tranferases + monooxygenases	[10,59,63-67]
Carbamates	acetylcholinesterase	$Ace1^R$ mutation		[10,59,63-67]
Neonicotinoids	nicotinic acetycholine receptor	not reported	monooxygenases	[68,69]
Spinosad	nicotinic acetycholine receptor	not reported		
Cyclodienes, Lindane, Bicyclic phosphates	GABA receptor	Rdl mutation	GS-transferases	[62,70]
Phenylpyrazoles	GABA receptor	Rdl mutation	GS-transferases	[62,70]
Avermectines	GABA receptor	undiscribed	monooxygenases + esterases	[62,70]
Insect Growth Regulators	Ecdysone agonist/disruptor or inhibitor of ATP synthase, chitin biosynthesis or lipid synthesis	no resistance reported against insect of public health importance		
Bacillus thuringiensis var. israelensis	microbial disruptors of insect midgut membranes	reported against Culex pipiens s.l. but non discribed		[71]

Historically, DDT was first introduced for mosquito control and malaria eradication programme in 1946. The first case of DDT resistance was reported in *An. sacharovi* in Greece in 1953 and was followed by dieldrin resistance in 1954 [15]. Onset of resistance was marked by deterioration in malaria control that has continued for more than 30 years with sporadic epidemics of disease [16]. Resistance in *An. sacharovi* has been later reported in Bulgaria, Lebanon, Iran, Iraq, and Syria [12]. Pronounced DDT resistance appeared in *An. stephensi* in Iran and Iraq when full scale house spraying operations began in 1957 and dieldrin resistance appeared three years later. In India, house-spraying of DDT and Lindane (HCH) under the public health programme was introduced in the 1950s. Resistance of the main malaria vector *An. culicifacies* to dieldrin developed in 1958 [17] and resistance to DDT in 1959 [18], but the malaria control programme continued until 1965-1966 when both DDT and HCH failed to control outbreaks of malaria [19]. As a result, malathion was introduced in some areas in 1969 with some success but *An. culicifacies* rapidly developed resistance by 1973 [20]. Malathion resistance resulted in colossal epidemics of malaria in 1975 with 4 million cases reported as compared with 125,000 in 1965. The experience in Pakistan was similar with DDT resistance appearing in 1963. The importance of the resistance was not recognized until outbreaks of malaria began in 1969 and neither DDT nor HCH was effective. By 1975, malaria cases were reported in Pakistan to number 100 million as compared to 9,500 in 1961 [12]. DDT resistance in *An. culicifacies* was reported in Sri Lanka in 1968 resulting in a severe epidemic of malaria [21]. This vector is now resistant to DDT, dieldrin, organophosphates, carbamates and pyrethroids [11].

Similar trend was noted in Central America and the Caribbean. Dieldrin spraying against *An. albimanus*begun in 1956 and widespread resistance appeared in 1958 [12]. A return to DDT spraying produced generalized resistance by 1960 [18]. The carbamate propoxur was employed in El Salvador, Guatemala, Honduras and Nicaragua in 1970 and resistance developed by 1974. *An albimanus* now exhibits multiple resistances to DDT, dieldrin, lindane and other chemical recently used in public health [22].

Much less information is however available for South East Asian malaria vectors, most probably because resistance monitoring was not carried out in routine before the 80s. In Vietnam, DDT resistance was found in 1989 in *An. epiroticus* of the Sundaicus Complex and is still occurring [23]. From 1990 till 2000, pyrethroid resistance was almost absent in all tested species except in some populations of *An. vagus* and *An. minimus s.l.* [24]. In Thailand, no evidence of insecticide resistance in malaria vectors was present before 1985 [25]. In 1986, development of physiological resistance to DDT was detected in *An. aconitus* from the north where DDT was commonly used for malaria control. One year later, DDT resistance was found in field collected mosquitoes of *An. philippinenis*, *An. nivipes* and *An. aconitus*from the same northern region. Between 1990 and 1997, DDT resistance has been detected in the three primary malaria vectors *An. dirus s.l.*, *An. minimus s.l.*

and *An. maculatus s.l.* and permethrin resistance was suggested in a population of *An. minimus s.l.* from northern Thailand, based however on a lower discriminative dosage (0.25%) of permethrin than that used today [25].

In Africa, resistance was initially found in *An. gambiae* in Bobo Dioulasso by 1967 (Burkina Faso), hence less than 7 years after the end of DDT use for malaria control [12]. DDT resistance was found in neighbouring countries including Cote d'Ivoire, Nigeria and Mali [26] and was then reported in most of Central and East African countries [27]. Strong association was observed between the level of DDT resistance in malaria vectors and the amount of DDT use for cotton protection [28]. Regarding BHC/dieldrin, the first cases of resistance were reported in Nigeria in 1954 hence only few months after the introduction of this molecule for malaria control. Initially found in very limited geographical areas, dieldrin resistance has spread in areas free of any insecticide treatments [29]. Few years later, resistance was reported in Bobo-Dioulasso and Cote d'Ivoire [30]. Today, resistance to BHC/dieldrin is still widespread in wild field anopheline populations despite its abandon in public health for many decades [31]. As for DDT, dieldrin resistance in malaria vectors arose and persisted from intensive use of pesticide for agricultural practices and in some specific settings due to public health programmes [32,33].

After the 80s, DDT has been more or less abandoned worldwide and replaced by organophosphate (OP), pyrethroids and, to lesser extent, carbamates. However, insecticide resistance continued to be a problem, and vector control operations were affected, particularly in India, Africa and Latin America, by extensive use of agricultural pesticides. OP resistance, either in the form of broad-spectrum OP resistance or malathion-specific resistance was found in the major malaria vector species worldwide [12]. Pyrethroids were introduced in late 70s in public health and increasingly used in the 90s; however, cases of resistance were rapidly reported in the main malaria vectors worldwide including *An. albimanus* [34], *An. darlingi*[35], *An. culicifacies* [36], *An. stephensi* [37], *An. gambiae* [38], *An. funestus* [39] and *An minimus* [40]. Despite a sporadic use (compared to DDT and pyrethroids), resistance to carbamates was earlier reported in several mosquito species including *An. albimanus* [41], *An. atroparvus* [42], *An. sacharovi* [5] and*An. gambiae* [43]. Carbamate resistance is now spreading in malaria vectors especially in West Africa where it has been reported in Cote d'Ivoire [44], Burkina Faso [45, 46], Benin [47] and Nigeria [48]. Increased level of carbamate resistance in African mosquito populations is worrying for malaria control because these chemicals are increasingly used in replacement to pyrethroids for Indoor Residual Spraying (IRS) [49].

It is obvious that insecticide resistance in malaria vectors is increasing worldwide due to the increasing selection pressure on mosquito populations caused by the presence of urban, domestic and/or agricultural pollutants in the environment [50]. Transversal and longitudinal monitoring surveys are essential to address the spatio-temporal changes in resistance (dynamic) and to design appropriate strategies for a better control of resistant malaria vector populations worldwide.

RESISTANCE MECHANISMS

The various mechanisms that enable insects to resist the action of insecticides can be grouped into four distinct categories including metabolic resistance, target-site resistance, reduce penetration and behavioral avoidance. These mechanisms that are shown in the figure 1 are briefly described in the following sections.

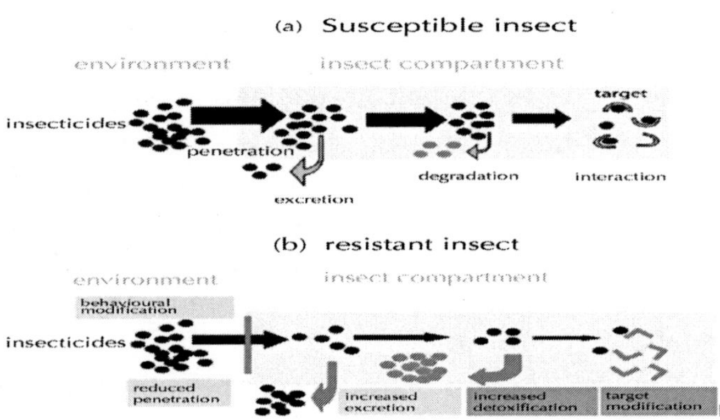

Figure 1: Scheme of potential behavioral and physiological changes associated with insecticide resistance in malaria vectors; (a) susceptible insect; (b) resistant insect *(*source *; see [14]).*

Metabolic Resistance

Metabolic resistance is the most common resistance mechanism that occurs in insects. This mechanism is based on the enzyme systems which all insects possess to help them to detoxify naturally occurring xenobiotics/insecticides. It is commonly accepted that insect detoxification systems derived from the plant-insect evolutionary arm race and several insect detoxification enzymes have been associated to the detoxification of plant toxins and all types of chemicals, including insecticides [51]. Over-expression of enzymes capable of detoxifying insecticides or amino acid substitutions within these enzymes, which alter the affinity of the enzyme for the insecticide, can result in high levels of insecticide resistance (see [52] for review). Increased expression of the genes encoding the major xenobiotic metabolizing enzymes is the most common cause of insecticide resistance in mosquitoes. Over expression of detoxyfing enzymes can occur as the result of gene amplification (e.g. duplication) or due to changes in either trans-acting regulator elements or in the promoter region of the gene [5, 53, 54]. The consequence is a significant increase of enzyme production in resistant insects that enables them to metabolize or degrade insecticides before they are able to exert a toxic effect. Three categories of enzymes, namely esterases, P450s and glutathione-S-transferases are known to confer resistance to insecticides in insect pest such as malaria vectors. These large enzyme families contain multiple enzymes with broad overlapping substrate specificities, and one member of the family might be capable of metabolizing limited number of insecticides. Similarly, the level of resistance conferred can vary from low to very high and may differ from compound to compound. Metabolic resistance mechanisms have been identified in mosquito populations for all major classes of insecticides currently used for vector control, including organochlorine, organophosphates, carbamates, and pyrethroids.

Esterases. One of the most common metabolic resistance mechanisms is that of elevated levels, or activity, of esterase enzymes which hydrolyze ester bonds or sequester insecticides. A striking example comes from studies on *Culex quinquefasciatus* that resist to a broad range of OP insecticides. In this species, multiple copies of EST-genes was already found hence enabling it to overproduce this type of enzyme [55]. In contrast to the situation in *Culex*, a number of *Anopheles* species (ie *An. culicifacies*, *An. stephensi* and *An. arabiensis*) have a non-elevated esterase mechanism that confers resistance specifically to malathion through increased rates of metabolism. Malathion resistance in *Anopheles* spp was associated with an altered form of esterase that specifically metabolizes the molecule at a much faster rate than that in susceptible counterparts [56, 57]. Although Carboxylesterase (CCEs) have been mostly associated to organophosphate resistance in mosquitoes, their role in pyrethroid resistance is probable. Indeed, the ability of esterases to metabolize pyrethroids has been suggested in mosquitoes [58, 59] even if no specific mosquito CCE has yet been validated as a pyrethroid metabolizer [50]. Clearly much more information is needed on the esterase-mediated resistance in malaria vectors.

P450s. Cytochrome P450-dependent monooxygenases are an important and diverse family of enzymes involved in the metabolism of numerous endogenous and exogenous compounds. P450 belong to six families and increased transcription of genes belonging to the CYP4, CYP6, and CYP9 has been observed in various insecticide-resistant species from different taxa [60]. There is increasing number of reports demonstrating elevated P450 monooxygenase activities in insecticide-resistant mosquitoes, frequently in conjunction with altered activities of other enzymes. In most cases where a link between insecticide resistance and elevated P450 activity has been shown, the CYP gene belongs to the CYP6 family. Since the publication of the *An. gambiae* genome [61], P450s were extensively studied in the primary malaria vector in Africa. A total of 111 P450 enzymes were identified [62] and, as in other insects, only a small number of these enzymes are capable of detoxifying insecticides. However, higher activity of enzymes and/or expression of detoxification genes in insecticide resistant colonies do not necessarily correlate with insecticide resistance. For example, some authors have shown elevated transcript levels of an adult-specific CYP6 P450 gene, CYP6Z1, in pyrethroid-resistant strain *of An. gambiae* [63, 64] and *An. funestus* [65]. Further validation studies conducted in *An. gambiae* showed that cyp6z1 was however not capable to metabolize pyrethroids but was capable to metabolize DDT [66]. Another study showed that CYP6z2 displays broad substrate specificity, which may be associated with xenobiotics metabolism and detoxification [67]. Despite, CYP6Z2 being able to bind to permethrin and cypermethrin, this gene does not metabolise any of these insecticides. Microarray-based approaches have lately identified three new "candidate" P450 genes that were found to be repeatedly over-produced in pyrethroid resistant populations of *An. gambiae*: CYP6M2, CYP6P3 and CYP6Z2 [68-70]. All of these genes encode for enzymes that are able to bind to type I and type II pyrethroids but only CYP6P3 and CYP6M2 showed to metabolize the insecticides [10, 71]. More recently, some authors demonstrated that CYP6M2 is also capable of metabolizing the organochlorine insecticide DDT in *An. gambiae*, hence demonstrating the first evidence for a metabolic cross-resistance in malaria vectors [9]. Interestingly the putative ortholog of*An. gambiae* CYP6P3, CYP6P9, as being the prime candidate for conferring pyrethroid resistance, have been identified in *An. funestus* [72, 73] but only the CYP6P9 showed to metabolize types I and II pyrethroids [74]. Recent works showed that over-production of CYP6P9 in *An. funestus* result from gene duplication [72]. In *An. minimus* mosquito, CYP6AA3 and CYP6P7 were up-regulated in pyrethroid-resistance population of Thailand [75] and seem to possess activities toward pyrethroid degradation [76,77].

Glutathione S-transferases (GSTs). Glutathione transferases (GSTs) are multifunctional enzymes involved in the detoxification of many endogenous and xenobiotic compounds. Conjugation of Glutathione (GSH) to such organic molecules enhances solubility, thus facilitating their eventual elimination [78]. Elevated GST activity has been implicated in resistance to at least four classes of insecticides in insects. Higher enzyme activity is usually due to an increase in the amount of one or more GST enzymes, either as a result of gene amplification or more commonly through increases in transcriptional rate, rather than qualitative changes in individual enzymes [52]. At least six classes of insect GSTs have been identified in *An. gambiae* [79], found in several large clusters on all three chromosomes. The Delta and Epsilon classes found exclusively in insects are the largest classes of insect GSTs. Members of both classes have been implicated in resistance to all the major classes of insecticide. The primary role of GSTs in mosquito insecticide resistance is in the metabolism of DDT to DDE (non toxic products), although they also have a secondary role in organophosphate resistance [80]. GST-based DDT resistance is common in a number of anopheline species including *An. gambiae* [81-83], reflecting the heavy use of this insecticide for malaria control over several decades. Molecular biology and *in vitro* expression studies showed that*aggst3-2* was over expressed in resistant strain of *An. gambiae* and that recombinant *aggst3-2* was very efficient at metabolizing DDT [84]. Most studies of GSTs suggested that regulation occurs at the transcriptional level. Several regulatory elements have been identified in the promoter regions of GSTs that may mediate their induction but the significance of these findings is unclear. Genetic mapping of the major genes controlling GST-based DDT resistance in *An. gambiae* provided however evidence for a trans-acting regulator [84], although in this species, mutations in promoter elements of the Epsilon GST cluster are also associated with resistance [81]. It has been suggested that GSTs may play a role in pyrethroid resistance by detoxifying lipid peroxidation products induced by pyrethroids and/or by protecting from insecticide exposure induced oxidative stress [85]. Furthermore, GST might confer secondary role in pyrethroid resistance by sequestering the insecticide hence reducing the total *in vivo*concentration of insecticide [86].

Despite the great advance obtained recently in the identification of the role of detoxifying enzymes in insecticide resistance, force is to note that the function of >90% of metabolic genes is still unknown. Although only a limited number of resistance mechanisms have been implicated to date, the diversity within enzyme families involved in metabolic resistance is likely to contribute substantially to resistance to many insecticide classes. Further functional genomics and post-genomic technology are needed to reveal the contributions of hitherto unsuspected enzymes in insecticide metabolism and/or sequestration and to identify the causal mutations associated with metabolic resistance in mosquitoes. The contribution that these enzymes make towards various insecticide resistance phenotypes in malaria vectors is yet to be elucidated.

Target-Site Resistance

The second most common resistance mechanism encountered in insects is target-site resistance. Insecticides generally act at a specific site within the insect, typically within the nervous system (e.g. OP, carbamate, DDT and pyrethroid insecticides). The site of action can be modified in resistant strains of insects such that the insecticide no longer binds effectively. Reduce sensitivity of the target receptors to insecticide results from non-silent point mutations in the gene encoding the protein. For example, the target site for OP and carbamate insecticides is acetylcholinesterase (AChE) in the nerve cell synapses. Several mutations in the gene encoding for an acetylcholinesterase have been found in insects" [87] which result in reduced sensitivity to inhibition of the enzyme by these insecticides [88, 89]. In malaria vectors, the G119S mutation (i.e. glycine to serine substitution at position 119) responsible for carbamate and OP resistance has been reported in *An. gambiae* and *An. albimanus,* essentially at the heterozygous state [90]. Recent sequence analysis of some resistant mosquitoes collected in Benin revealed the presence of a duplication of the *ace-1* gene in both A*n*. *A. gambiae* M and S forms [91]. In addition, mutations at a single codon (position 302) in the Rdl (resistance to dieldrin) gene encoding one receptor subunit, from an alanine residue to a serine (or more rarely to a glycine), have been documented in dieldrin-resistant insect species [92] including the malaria vectors *An. stephensi* [93], *An. gambiae s.l.* [94] and *An. funestus* [31]. Similarly, mutations in the amino acid sequence in the voltage-gated sodium channels of nerve cell membranes leads to a reduction in the sensitivity of the channels to the binding of DDT and pyrethroid insecticides [95]. Alterations in the target site that cause resistance to insecticides are often referred to as knockdown resistance (*kdr*) in reference to the ability of insects with these alleles to withstand prolonged exposure to insecticides without being 'knocked-down" [96]. One of the most common amino acid replacements associated with pyrethroid resistance in malaria vectors is a substitution of the leucine residue found at codon 1014 with either phenylalanine (1014F) [97] or serine (1014S) [98] in the Voltage-Gated Sodium Channel (VGSC). Interestingly, residue 1014 does not appear to interact directly with the insecticide but is predicted to alter channel activation kinetics [99]. Note that a *de novo* mutation (N1575Y) recently emerged within domains III-IV of voltage gate sodium channel in pyrethroid resistant populations of *An. gambiae* and seems to occurs only in a single long-range haplotype, also bearing 1014F allele [100]. It has been suggested that the N1575Y mutation may compensates for deleterious fitness effects of 1014F and/or confers additional resistance to pyrethroid insecticides.

Reduced Penetration

Modifications in the insect cuticle or digestive tract linings that prevent or slow the absorption or penetration of insecticides can be found is some resistant insects. This resistance mechanism is not specific and can affect a broad range of insecticides. Reduced uptake of insecticide, often referred to as cuticular resistance, is frequently described as a minor resistance mechanism. Certainly for pests where the major route of insecticide delivery is via ingestion, this is likely to be the case. However, for malaria control, where insecticides are typically delivered on bed nets or on wall surfaces, uptake of insecticides is primarily through the appendages. An increase in the thickness of the tarsal cuticle, or a reduction in its permeability to lipophilic insecticides, could have a major impact on the bioavailability of an insecticide*in vivo*. Examples of reduced-penetration mechanisms are however limited; cuticular resistance was reported for the domestic Fly *Musca domestica* [101] and the lymphatic filariasis vector *Culex quinquefasciatus* [102]. In *Anopheles*, microarrays studies have recently identified two genes, cplcg3 and cplcg4, encoding cuticular proteins that were upregulated in pyrethroid resistant strains from four populations and two different species (i.e. *An. gambiae* and *An. stephensi*) ([69, 103, 104]. Recently, measures of mean cuticle thickness in a laboratory strain of *An. funestus* using scanning electron microscopy (SEM) showed that the mean cuticle thickness was significantly greater in pyrethroid tolerant mosquitoes than their susceptible counterparts [105]. Clearly much more work is required in order to identify the significance of cuticular resistance in phenotypic resistance.

Behavioural Resistance

Insecticide resistance in mosquitoes is not always based on biochemical mechanisms such as metabolic detoxification or target site mutations, but may also be conferred by behavioural changes in response to prolonged exposure to an insecticide. Behavioural resistance does not have the same "importance" as physiological resistance but may be considered to be a contributing factor, leading to the avoidance of lethal doses of an insecticide [106, 107]. For example, the first study on the irritant effect of DDT residual deposits was conducted using

Anopheles quadrimaculatus where females were found to be irritated shortly after making contact with the treated surfaces resulting in a rapid escape response from a treated house prior to taking a blood meal [108]. This type of response can be further divided into direct contact excitation (sometimes referred to as 'irritancy') and non-contact spatial repellency that is used when insects move away from the insecticide-treated area before making direct contact [106, 109]. Examples of behavioral resistance or avoidance are few. Change in vector composition (i.e. switch from*An. minimus* to *An. harrisoni*) has been observed following implementation of ITNs in a village form central Vietnam [110]. With regard to *An. funestus*, recent findings showed a shift from indoor to outdoor biting preferences in Tanzania in relation to increasing coverage of pyrethroid-impregnated net [111]. Significant changes in the host-seeking behavior of the *An. funestus* population was confirmed in Benin (West Africa) where scaling up of LLINs at community level induced a change from night biting to early-morning biting behaviour [112]. It is unclear however whether adaptation of malaria vectors species to insecticidal based vector control interventions such as LLIN may result from a phenotypic plasticity or from selected behavioral traits (see Durnez & Coosemans for details).

METHOD TO DETECT INSECTICIDE RESISTANCE

Currently most resistance monitoring is dependent on bioassays, using fixed insecticide concentrations and exposure times, and the data is reported as percentage mortality and/or Knock Down (KD) effect. The World Health Organisation (WHO) has defined diagnostic doses (i.e. twice the dosage that killed 100% susceptible mosquitoes of a given species) for most insecticides used in malaria control and produces susceptibility test kits consisting of exposure chambers and insecticide treated filter papers [113-115]. Although simple to perform, these diagnostic dose assays provide limited information and several alternative methods for detecting resistance are available (Table 2). These alternative assays generally detect specific resistance mechanisms, and should always be performed as an addition, not a substitute, to bioassays, to avoid the risk that unknown resistance mechanisms go undetected. It should be noted that none of the current methods listed in Table 2 are suitable for detecting cuticular and/or behavioural resistance. Regular monitoring for insecticide resistance is essential in order to react proactively to prevent insecticide resistance from compromising control. If the frequency of resistance alleles is going to build up unchecked, resistance may eventually become 'fixed' in the populations. Once resistance reaches very high levels, strategies to restore susceptibility are unlikely to be effective.

Table 2: Methods for detecting insecticide resistance (source: see [6])

Method	**Advantages**	**Disadvantages**
Bioassays using WHO defined diagnostic doses of insecticide	Standardized, simple to perform, detect resistance regardless of mechanism	Lack sensitivity and provide no information about level and type of resistance (except when using with synergists), to be done on live mosquitoes
Dose response bioassays	Provides data on level of resistance in population, regardless of mechanism	Require large numbers of alive mosquitoes, and data from different groups not readily comparable
Biochemical assays to detect activity of enzymes associated with insecticide resistance	Provides information on specific mechanisms responsible for resistance	Requires cold chain. Not available for all resistance mechanisms, sensitivity and specificity issues for some assays (e.g. GST)
Molecular assays to detect resistant alleles	Very sensitive. Can detect recessive alleles and therefore provide an 'early warning' of future resistance.	Requires specialized and costly equipment. Only available for a limited number of resistance mechanisms.

Bioassays

Guidelines for test procedures and interpretation of results are available from the WHOPES[3] - (see http://www.who.int/whopes/resistance/en/). It is important that the mosquitoes used for the bioassays are standardized for age, sex and physiological status as all of these can affect the outcome of the tests. Typically either adults raised from isofemale lines or F1 progeny from field collected blood fed females are used. The limitations and advantages of these two alternatives have recently been discussed [116].

These diagnostic dose assays are simple to perform and provide standardized data sets that, assuming the guidelines are followed, can be readily compared to identify temporal and/or geographical variations in the resistant status of malaria vector populations. However, it is important to recognize some of the limitations of these susceptibility tests. As only a single concentration of insecticide is used, the results do not provide any information about the level of resistance in a population. For example if 50 % of population A and 20 % of population B were killed after exposure to the diagnostic dose of permethrin, it cannot be concluded that population B is more resistant than population A. The results only indicate that both populations are resistant (according to WHO definitions if there is < 80 % mortality, the population is defined as resistant) and that, subject to tests of significance, there is a higher frequency of resistant individuals in population B than in A. Dose response assays would be needed to compare the levels of resistance in two populations (e.g. by measuring the Resistant Ratios and their 95% confidence intervals). For pyrethroids, median knock down time (MKDT) is also a useful quantifiable variable [117]. Similarly, the results of these tests cannot be used to compare the levels of resistance to two different insecticides. If 50 % mortality was observed after exposure to the diagnostic dose of permethrin (0.75 %) whereas mortality was 70% after exposure to the diagnostic dose of deltamethrin (0.05%), it is not correct to state that the population is more resistant to permethrin than deltamethrin. Again, all that can be stated is that the population is resistant to both insecticides.

Partly due to the limitations of the diagnostic dose assays described above and partly due to the difficulties that are sometimes incurred in obtaining a regular supply of the insecticide impregnated papers from WHO, an alternative bioassay methodology has been developed [118] and is being adopted by some monitoring programmes. This method, known as the CDC bottle bioassay, uses glass bottles coated with a known concentration of insecticide. As these test kits are assembled in the users own laboratory, the concentration of insecticide can be readily adjusted enabling dose response curves to be developed to compare two or more strains. A caveat to this is that the flexibility, and the potential variation in the insecticide grade used in the tests, impairs comparison of results between two separate studies.

Both WHO diagnostic doses and CDC bottle bioassays can be modified to incorporate synergists. Synergists such as piperonyl butoxide, that block the activity of two major detoxification enzyme families, can be used to explore the role of different resistance mechanisms. If resistance is due to increased metabolism, exposure to an appropriate synergist prior to insecticide bioassays should increase the level of mortality observed.

Biochemical Tests

Biochemical tests to detect alterations in activities of enzyme families associated with insecticide resistance have been available for over two decades and are sometimes used in combination with insecticide bioassays [119]. These assays employ model substrates to record the overall activity of glutathione transferases, carboxylesterases or cytochrome P450s in individual insects. Biochemical assays are also available to detect target site resistance to organophosphate and carbamate insecticides caused by insensitive acetylcholinesterase (AChE). The enzymatic reaction produces a colour change that is generally visible to the naked eye and hence these assays do not require access to expensive equipment (spectrophotometer is appropriate). However, it is important that the mosquitoes are kept on ice from the point of collection to the performance of the assay and this can often pose logistical challenges. Furthermore, there are sensitivity and specificity issues that limit the utility of some of these assays. For example, with over 100 different cytochrome P450 enzymes in malaria vectors, an assay that measures the total level or activity of this enzyme family may not have the sensitivity to detect over production of the single or small number of P450 enzymes that are thought to be involved in pyrethroid metabolism. This may explain the lack of significant correlation observed in many studies between cytochrome P450 activity and bioassay mortality results [120, 121]. In addition not all members of the enzyme family will have the same affinity for the model substrates used in these assays (e.g. CDNB (1-chloro 2-4, dinitrobenzene) is the substrate typically used to assess glutathione transferase activity but the Epsilon class of GSTs which are responsible for DDT resistance have relatively low activity with this substrate). In order to incorporate data from resistance monitoring into evidence based decisions on appropriate insecticide based interventions for malaria control, it is clearly essential that the data is both reliable and accessible. Although guidelines for conducting the various assays exist, there is little consensus on the number of sites and frequency with which resistance monitoring should occur [122]. It is clear that resistance is a dynamic trait, and wide fluctuations in resistance levels throughout the malaria transmission season have been reported [116, 123, 124]. Resistance can also be very focal, particularly when vector composition differs between sites [125], hence a minimum number of sampling sites should be established, taking into account patterns of vector distribution and insecticide usage. The WHO/AFRO African Network for Vector Resistance was established in 2000 and amongst its objectives was the important goal of improving the dissemination of resistance data. Accordingly a database was established to store the results of resistance monitoring activities by the African Network for Vector Resistance (ANVR) members but until recently, this database was not readily accessible by outside users. The recent establishment of new data base (see section 6), as an online centralized resource for collating data on insecticide resistance in disease vectors and the integration of this with the ANVR database, will hopefully ensure that both published and unpublished data on resistance in malaria vectors are more readily available to all interested parties.

Molecular Tests

A multitude of molecular assays have been developed to detect *kdr* alleles in malaria mosquitoes, several of which were recently compared in a study by Bass et al (see [126]). These are routinely used by research laboratories monitoring for insecticide resistance and are gradually being incorporated into some national malaria control resistance monitoring programmes. Unfortunately, despite the recent identification of the key enzymes responsible for metabolic resistance to pyrethroids in *An gambiae* and *An funestus*, there are currently no simple DNA based assays to detect these resistance mechanisms. Detection of these genes is presently dependent on RNA based approaches using relatively sophisticated equipment (e.g. RT-qPCR). Assays to detect the genetic mutation(s) responsible for the resistance phenotype in individual insects can provide an early warning of the emergence of resistance which may not have been detectable by bioassays that can only record the population response. The presence of a single individual with an allele known to confer resistance should be cause for concern as experience dictates that resistance can spread very rapidly in a population unless the selection pressure is eased and/or the genetic cost associated with the resistant allele is high.

Conversely, a negative result from a molecular assay should not lead to complacency. As discussed above, molecular assays are presently only available for target site resistance and the failure to detect *kdr* clearly cannot be interpreted as an absence of resistance in a population. Hence molecular assays should be seen as a complement rather than a substitute for bioassays.

CURRENT DISTRIBUTION OF INSECTICIDE RESISTANCE

Insecticide resistance has been reported in the main malaria vectors worldwide. Resistance is however not uniformly distributed among vector species and can greatly differ from one village, province, country, region and continent to another. Unfortunately, the highest levels of insecticide resistance were reported in Africa where malaria burden is still the highest in the world [1]. Resistance to pyrethroids, the gold standard insecticides used for LLIN and IRS will be extensively discussed in the present chapter as it remains a real and ever-present danger to future success of malaria vector control. Note that more information on the distribution of insecticide resistance in malaria vectors can be found in *Anobase,*http://anobase.vectorbase.org/ir/; *MARA* http://www.mara.org.za; *Arthropod Pesticide Resistance Database*, http://www.pesticideresistance.org.; and IR mapper, http://www.irmapper.com.

Africa

Although the occurrence of insecticide resistance in malaria vectors in Africa is not a "new" event (see section 2.), the speed at which pyrethroid-resistance recently evolved in field populations is worrying as it may jeopardize the current malaria vector control initiatives carried out in the continent. As shown infigure 2, pyrethroid resistance in *Anopheles* sp. is widespread but not uniformly distributed among the different countries. In the 49 African countries that have been investigated (see [6] for details), 15 did not report any data on resistance in the last 10 years i.e. Algeria, Botswana, Democratic Republic of Congo, Djibouti, Sierra Leone, Lesotho, Liberia, Libya, Maurice, Mauritania, Namibia, Rwanda, Somalia, Swaziland, Tunisia. If a lot of data has been generated in West Africa (as far as *An. gambiae s.l.* is concerned), a lack of information is globally observed in Central, Eastern and Austral Africa. It is obvious that the frequent conflicts that has occurred in the last decades in some African countries has rendered difficult the conduct of routine monitoring surveys by NMCP, International Organisation (WHO/ANVR) and/or research institutions.

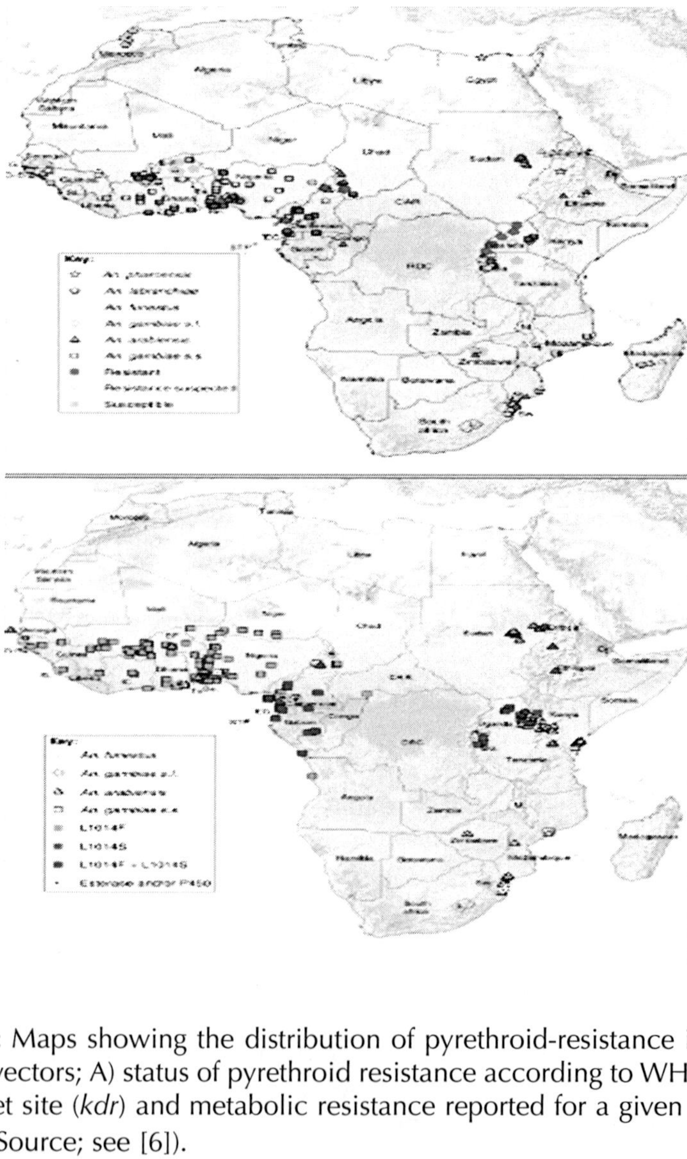

Figure 2: Maps showing the distribution of pyrethroid-resistance in African malaria vectors; A) status of pyrethroid resistance according to WHO criteria ; B) Target site (*kdr*) and metabolic resistance reported for a given mosquito species (Source; see [6]).

Globally, pyrethroid resistance is high in *An. gambiae s.l.* in West Africa including Benin [127], Burkina Faso [128], Guinea Konakry ([129], Ghana [130], Mali [131], Niger [132], Nigeria [133] and Cote d'Ivoire [134]). In this region, pyrethroid resistance is predominant in *An. gambiae s.s.*, compared to *An. arabiensis*. Surprisingly, susceptibility to pyrethroids (permethrin and/or deltamethrin) was reported in*An. gambiae* s.l. in Guinea Bissau [135] despite the presence of the L1014F mutation. In Central Africa, pyrethroid resistance/tolerance is widespread in *An. gambiae s.l* in Cameroon [136-138], Chad [116,139], Gabon [140, 141], Equatorial Guinea [8] and Sudan [142, 143]. In Chad, North Cameroon and Sudan, pyrethroid resistance is present essentially in *An. arabiensis,* which is consistent with the higher prevalence of this mosquito species in more arid areas with higher mean annual temperatures [144]. In East and Austral Africa, *An. gambiae* and *An. arabiensis* populations are mostly susceptible to pyrethroids in Tanzania [145, 146], Mozambique [147] and Madagascar [148], but highly resistant in eastern Uganda [149, 150], Ethiopia [151], Kenya [152, 153], Zambia [154], South Africa [155] and the Gwave Region of Zimbabwe [120]. Regarding *An. funestus,* most of the literature reporting pyrethroid resistance comes from South Africa [39, 156] and Mozambique [157-159], most probably because *An. funestus* is the main malaria vector in these countries. The data available in other African countries is very limited in partly due to the difficulty to colonize *An. funestus* species in laboratory. Pyrethroid resistance/tolerance was detected in Malawi [160, 161] and suspected in Obusi and Kassena-Nankana Regions from Ghana [121, 162] and Benin [163] whereas full susceptibility to permethrin and deltamethrin was found in Burkina Faso [164] and Tanzania [145]. There is a lack of information on secondary vectors e.g. *An. moucheti* and *An. nili* which can play important role in malaria transmission in specific settings (e.g. Cameroon, Congo, Côte d'Ivoire). Regarding other vectors species, fullsusceptibility to pyrethroids has been reported in *An. labranchiae* in Morocco [165] and in *An. pharoensis* in Egypt [166] and Ethiopia [167].

In Africa, the L1014F mutation is widespread (figure 2) and predominant in the molecular S form compared to the M form, except in Benin [168], Guinea Equatorial [8] and Niger [132]. Some authors suggested that the *kdr* alleles may have arisen from at least four independent mutation events in the *An. gambiae* S-form [169].

Regarding the M form, it is not clear whether the *kdr* mutation resulted from an introgression from the S form only [170, 171] and/or from independent mutation events, has recently suggested for Bioko Island [172]. The second mutation, a leucine–serine substitution at the same codon (L1014S), was identified first in a colony of *An. gambiae s.l.* from Kenya [98]. This substitution has been lately reported in Burundi [173], Cameroon [136, 138], Gabon [174], Equatorial Guinea [175], Uganda [176], Republic of Congo [177] and Angola [140], mainly in co-occurrence with the 1014F *kdr* allele. Although some authors have reported that the 1014S allele may confer lower level of pyrethroid resistance than the 1014F allele [178], its spread from eastern to central Africa and more recently to West Africa [124, 179] suggest a survival advantage of mosquitoes sharing this mutation in presence of pyrethroids. So far, the L1014S substitution has always been detected in the S molecular form [180] but recent findings showed the occurrence of the 1014S allele in the M form in Equatorial Guinea [181] and Cameroon [182]. In these two countries, the 1014S allele was present at very low frequencies, alone or associated with the L1014F allele. It is currently impossible to know whether the *kdr* alleles have arisen first in Cameroon or Equatorial Guinea. The higher frequency of the 1014S allele in the S form compared with the M form could either be attributed to an introgression from the S taxon or to a *de novo* mutation. Regarding the sister taxa *An. arabiensis*, both of these mutations were reported in Western [124], Central [183] and Eastern Africa [184]. Interestingly, a new *kdr* mutation (N1575Y) occurring within domains III-IV of voltage gate sodium channel was found in both S and M molecular forms of *An. gambiae* and occurs upon a 1014F haplotypic background only [100]. Additive resistance of *1575Y* was demonstrated for permethrin and DDT in both molecular forms of *An. gambiae*. The prevalence of the 1575Y mutation has increased in West Africa in the last years hence indicating that the 1014F-1575Y haplotype is under strong selection pressure (Djégbé pers. com). It is possible that besides the 1014F/1014S *kdr* mutation, other mutations in the para-type sodium channel gene might be needed for mosquitoes to survive after exposure to a discriminating concentration of an insecticide. Further investigation is needed to better address the distribution and the role of the N1575Y mutation in pyrethroid resistance as well as to assess the fitness benefits conferred by this allele on the L1014F mutation in malaria vectors.

Beyond the spread of *kdr* alleles, metabolic-based resistance due to detoxifying enzymes namely oxidase, the GST (epsilon) and CCE families have expanded in African malaria vectors. In *An. gambiae s.l.*metabolic resistance involving increased levels of P450 has been reported at least in Kenya [185], Cameroon [186], Benin [69], Nigeria [69], Ghana [70], Mozambique [147], South Africa [187] and Zimbabwe [120]. Up to now, only genes encoding CYP6P3 and CYP6M2 P450 enzymes have been clearly involved in cellular mechanisms known to metabolize deltamethrin and permethrin [10, 71].These genes were found over-expressed in pyrethroid-resistant *An. gambiae* populations from Benin, Nigeria and Ghana [69, 70], mainly in co-association with the *kdr* L1014F allele. In *An. funestus*, pyrethroid resistance involving increased activity of P450 monooxygenase and/or GST was demonstrated in South Africa [157], Mozambique [188] and Malawi [161].

To conclude, the immense challenge in Africa will be not to manage and control *kdr*-resistant mosquitoes only but to deal with the development of "multiple resistant" populations that could resist to different class of insecticides used in public health. One other issue is the occurrence and development of carbamate resistance in some countries (eg Benin, Nigeria) where this chemical class is in use for IRS through the PMI programme [47, 48]. The spread of carbamate resistance in malaria vectors in Africa is worrying for insecticide resistance management and alternative insecticides, and innovative strategies are urgently needed to better reduce the vectorial capacity of mosquitoes and hence effectively reduce the burden of malaria in the region. Resistance management strategy for malaria control is discussed in section 8.

South-East Asia and India

The South East Asia Region (SEAR) that account for 13% of the total malaria cases worldwide (2ndposition after Africa) [1] is not spare of insecticide resistance in the main malaria vector species.

In the Mekong region, cross-country monitoring of insecticide resistance has been conducted through the MALVECASIA network (http://www.itg.be/malvecasia/) to help MCPs in the choice of insecticide to use at regional level. Large differences in insecticide resistance status were observed among species and countries. *Anopheles dirus s.s.*, the main vector in forested malaria foci, was mainly susceptible to permethrin except in central Vietnam where it showed possible resistance to type II pyrethroids [23].*Anopheles minimus* s.l. populations were found resistant / tolerant in Vietnam and northern Thailand [189] but almost susceptible in Cambodia and Laos. No *kdr* mutation has been observed so far in these species [190] and pyrethroid resistance seems to result from increased detoxification by esterases and/or P450 monooxygenases [191]. Indeed, increased mRNA expression of two P450 genes, *CYP6P7* and*CYP6AA3*, suspected to metabolize some pyrethroids [76] have been reported in a deltamethrin-resistant population of *An. minimus* in Thailand [75, 192].

Anopheles epiroticus of the Sundaicus Complex showed to be highly resistant to all pyrethroids in the Mekong Delta [23] but susceptible to DDT, except near Ho Chi Minh City. DDT and pyrethroid-resistant populations of *An. subpictus* were reported in Vietnam and Cambodia. Biochemical assays suggest an esterase-mediated pyrethroid detoxification in both *An. epiroticus* and *An. subpictus* whereas DDT resistance in *An. subpictus* might be conferred to a higher GST activity. In Vietnam and Cambodia, *An. vagus* and *An sinensis* showed various levels of pyrethroid resistance and sequence-analysis of the DIIS6 region of the VGSC revealed the presence of the 1014S *kdr* allele [193]. Pyrethroid resistant populations of *An. sinensis* were also reported in the Republic of Korea (ROK) [194] and in China [195]. In China, cypermethrin resistance in *An. sinensis* was associated with the presence of both 1014F and L1014C substitutions, whereas only 1014F and 1014S mutations were found in the ROK and Vietnam, respectively. In Indonesia, molecular analysis carried out in field mosquito samples revealed the presence of the 1014F allele in the four main malaria vectors i.e. *An. sundaicus*, *An. aconitus*, *An. subpictus* and*An. vagus* [196]. At the present time, it is difficult to speculate on the relative contribution of the *kdr*mutations *versus* metabolic detoxification on pyrethroid and DDT resistance in malaria vectors from the SEA region and more work are needed to establish a clear trend.

Insecticide resistance is known to be widespread in other part of Asia such as India. In this country, resistance has a long history (see section 2) and it represents a big challenge for malaria vector control. Among the *Anopheles* species, *An. culicifacies s.l.*, the major vector of malaria in most parts of the country, has developed strong resistance to pyrethroids [36], DDT [197, 198], dieldrin/HCH [199], and malathion [198]. The 1014F mutation, which generates the *kdr* phenotype was detected in pyrethroid and DDT resistant *An. culicifacies s.l.* populations sometimes in co-occurrence with the 1014S mutation [197]. Note that a novel mutation V1010L (resulting from G-to-T or -C transversions) in the VGSC was recently identified in Indian *An. culicifacies* and was tightly linked to 1014S substitution [197]. Elevated activities of GST seem to play also an important role in DDT-resistance in this mosquito species [82]. Similarly, strong level of pyrethroid resistance due to the presence of both 1014F and 1014S mutations was found in the urban malaria vector *An. stephensi* particularly in the Rajasthan District [200]. Other vectors that are reported to be resistant to pyrethroid, DDT and/or dieldrin/HCH in India are *An. annularis, An. subpictus* and *An. philippinensis* [201]. In contrast, *An. minimus* has still not showed pyrethroid and DDT resistance [202].

The same trend was noted in Sri Lanka where the main malaria vectors species, i.e. *An. culificifacies s.l.*and *An. subpictus* have developed DDT, pyrethroid and malathion resistance in several districts [203,204]. However, the main mechanisms associated with DDT and malathion resistance in *An. culicifacies s.l.* and *An. subpictus* are primarily metabolic and involve carboxylesterases (malathion) or monooxygenases and GSTs (for DDT) [205, 206]. An altered acetylcholinesterase conferring organophosphate resistance has been suspected in both vector species [205].

In the delta region of Bangladesh, the *An. sundaicus* malaria vector is fully susceptible to DDT but other malaria vectors such as *An. philippinensis, An. maculatus s.l.,* and *An. aconitus* have all developed resistance to DDT [207]. *Anopheles aconitus,* additionally, has been reported to be resistant to dieldrin/HCH. Bhutan records *An. maculatus s.l.* as resistant to DDT, but there is no record of its resistance to any other insecticides [207]. Two vectors of malaria in Nepal, *An. maculatus s.l.* and *An. aconitus,* also have developed resistance to DDT whereas only malathion resistance was reported in *An. stephensi* in Pakistan [208]. Finally, in Iran and Turkey, *An. stephensi* and *An. sacharovi* showed resistance to DDT and dieldrin but both species are mostly susceptible to pyrethroids [209-211].

Latin America

The countries of the Amazon Basin (Bolivia, Brazil, Colombia, Ecuador, Guyana, Peru, Surinam and Venezuela) carry the greatest burden of malaria in the Americas. The primary vectors of this disease in the Amazon basin are *An. darlingi* and *An. albimanus*. Surprisingly, much less data on insecticide resistance is available for these two mosquito species comparatively to African and/or Asian malaria vector species [212].

In Colombia, DDT resistance was reported in the late 80's in some populations of *An. darlingi* in the districts of Quibdó and close to the Atrato River [213, 214]. Successive insecticide susceptibility evaluations revealed resistance to pyrethroids in both *An. darlingi* and *An. albimanus* mainly in the Chocó State [215]. In *An. darlingi*, increased levels of both Multi function Oxidase (MFO) and Nonspecific Esterase (NSE) were reported in a deltamethrin and DDT-resistant population, hence suggesting a possible involvement of these detoxifying enzymes in cross resistance to DDT and deltamethrin [35]. Note that various levels of resistance to organophosphate and pyrethroids were also reported in the secondary malaria vector *An. nuneztovari* [216].

In neighboring countries, DDT, permethrin and deltamethrin resistance was found in laboratory colonized populations of *An. albimanus* from Guatemala, whereas full susceptibility was noted in field populations from El Salvador and Belize [217, 218]. The colonies from Guatemala showed significant increase in the specific activity of esterase and/or oxidase as measured by spectrophotometer suggesting their potential involvement in pyrethroid-resistance [34, 217]. In Peru, monitoring campaigns carried out since 2000 showed that *An. albimanus* was the only Anopheline species to exhibit pyrethroid-resistance [219].

In Mexico, high level of DDT resistance and low levels of resistance to organophosphate, carbamate and pyrethroid insecticides were detected in field populations of *An. albimanus* in Chiapas, prior to a large-scale resistance management project [220]. Biochemical assays revealed that the DDT resistance was caused by elevated levels of GST activity leading to increased rates of metabolism of DDT to DDE [22], whereas carbamate resistance was attributed to an altered acetylcholinesterase (AChE). More recent studies conducted in the southern Yucatan Peninsula showed high levels of DDT, deltamethrin and pirimiphos-methyl resistance in the *An. albimanus* populations tested [221]. Biochemical tests revealed elevated levels of GST, P450 and esterases activities that could be involved in DDT and pyrethroid-resistance. As for carbamate, pirimiphos-methyl resistance was strongly correlated with the presence of an insensitive acetylcholinesterase.

To our knowledge, it is the main "published" information available on the distribution, levels and mechanisms of resistance (i.e. accessible through Medline and pub med) in malaria vectors in Latin America. It is then essential to strengthen the capacity of all Latin America countries that suffering from malaria to make insecticide monitoring in routine to obtain much accurate information on the insecticide resistance situation in the malaria vectors. This will provide stake holders with useful information for the implementation of more effective and sustainable malaria control programmes in the region.

IMPACT OF PYRETHROID-RESISTANCE ON PROGRAMMATIC MALARIA CONTROL

Few operational reports exist that measure the impact of pyrethroid resistance on epidemiological outcomes of malaria, owing to body of factors that mislead the attributable component of resistance. Where tentative evidence is provided in most cases, the design of the study has been observational and the effect of confounding factors can never be excluded with confidence, making difficult the interpretation of data.

Most probably, the only clearest evidence of control failure being directly linked to pyrethroid resistance was reported from the borders of Mozambique and South Africa. In 1996, the malaria control programme in KwaZulu Natal switched from using DDT to deltamethrin for indoor spraying. Within four years, notified malaria cases had increased about four fold, *An. funestus* had re-appeared and was observable emerging alive from pyrethroid sprayed houses. Bioassays showed that this species was resistant to pyrethroids but susceptible to DDT [39]. A decision was taken to switch back to DDT spraying and, within the two years after this switch was made, *An. funestus* was no longer observed emerging alive from insecticide sprayed houses. The combination of DDT and antimalarial drugs in KwaZulu-Natal has resulted in a 91% decline in the malaria incidence rate [222, 223]. There is no doubt that that the emergence of pyrethroid resistance and the avoidance of its effects by switching to DDT, has been of major operational importance [224].

Additional evidence was brought on the island of Bioko on the West African Coast. A malaria control strategy based on IRS with lambdacyhalothrin was launched by the Bioko Island Malaria Control Project (BIMCP) funded by the Government of Equatorial Guinea and a consortium of private donors led by Marathon Oil Corporation. One round of IRS using the pyrethroid deltamethrin (K-Orthrine WP50, Bayer Crop Sciences, Isando, South Africa) failed to curtail an increase in the population density of *An. gambiae* M form because of evidence in the rise of the knock-down resistance (*kdr*) gene in this species [8]. The programme switched to carbamate insecticide before a substantial decline in the mosquito population, transmission index and malaria prevalence in children was seen. Nevertheless, in an observational study such as this, the possible contribution of other confounding factors to the failure of pyrethroid IRS cannot be overlooked so the direct consequence of the *kdr* frequency is unclear.

Another programmatic study was conducted in the highland provinces of Burundi. Between 2002 and 2005, a well targeted vector control programme (conducted in foot of valleys only) combining IRS with pyrethroids and/or PermaNet 1.0 LLINs was initiated in one of the most affected island provinces, Karuzi [225]. Initially, one round per year of pyrethroid-IRS was carried out in all human dwellings and cattle sheds before the seasonal increase in transmission. LLIN distribution preceded the first IRS round in the same year. The S-form of *An. gambiae* was the predominant vector species in Karuzi District and showed resistance to pyrethroids due to the *kdr* mutation. The entomological data showed that the intervention, overall, effectively reduced *Anopheles* density by 82% and malaria transmission was decreased by 90% despite high frequencies of the L1014S allele in the local *An. gambiae* population [173].

In a more recent observational study conducted in Malawi, the impact of pyrethroid resistance on operational malaria control has been assessed with more controversial evidence of resistance impacting pyrethroid-based vector control [161]. In this trial, pyrethroid-LLINs were distributed to communities in 2007 followed by a pilot campaign of IRS with lambdacyhalothrin supported by the President's Malaria Initiatives between 2008-2010 within districts. A series of sentinel sites were established during these periods to track the effect of the increase in pyrethroid resistance in the local malaria vectors (*An. gambiae* and *An. funestus*) and assess any impact on malaria transmission and prevalence of infection. Pyrethroid resistance had been selected over the 3 years of the programme in these two major malaria vectors with the resistance in the later vector (i.e; *An. funestus*) being metabolically-mediated and involving the up-regulation of two duplicated P450s. The selection of resistance over 3 years had however not triggered a major increase in parasite prevalence in Malawian children, but it may have reduced the benefit of introducing IRS alone in several districts [161]. The impact of this pyrethroid resistance on the ability of LLIN and IRS to reduce malaria infection in Malawi needs to be further elucidated.

Similarly, in the Dielmo Village of Senegal, a longitudinal study of inhabitants was carried out between January, 2007, and December, 2010 [226]. In July, 2008, deltamethrin-LLINs were provided to all villagers and asymptomatic carriage of malaria parasites was assessed from cross-sectional surveys. Overall, the incidence density of malaria attacks decreased from 5.45 per 100 person-months before LLINs distribution in 2007 to 0.41 by August 2010, but increased sharply back to 4.57 between September and December, 2010, i.e, in less than 3 years after the distribution of LLINs. Within the same time frame, the malaria vector became gradually resistant to pyrethroids and the prevalence of the 1014F*kdr* resistance allele increased from scratch, i.e. 8% in 2007 to 48% in 2010. Once again, these results should be considered with caution as the study was conducted in an unique village and the conclusions drawn could not be extrapolated or extended to Senegal or other areas of Western Africa. Moreover, the link between the slight rise of pyrethroid resistance and the rebound in malaria cases cannot be established with accuracy and such rebound could be due to other sources of factors totally independent of resistance.

Another recent study reports the presence of pyrethroid-resistance in malaria vectors *versus* the gain in current efforts to control malaria in the Zambia [154]. In line with the Global trend to improve malaria control efforts, a country wide campaign of Olyset Nets and PermaNets (LLIN) distribution was initiated in 1999 and indoor residual spraying with DDT or pyrethroids was reintroduced in 2000 in the country by the NMCP. In 2006, these efforts were strengthened by the PMI. Both major malaria vectors, *An. gambiae* and *An. funestus* were controlled effectively with the ITN and IRS programme in Zambia, maintaining a reduced disease transmission and burden, despite the discovery of DDT and pyrethroid resistance in the country.

There have been extensive randomized controlled trials (RCTs) (phase III) in part of Africa aiming at investigating the efficacy of ITNs for malaria prevention [227], but very few have assessed how pyrethroid resistance might affect the effectiveness of such intervention. RCTs entail a set of communities randomly divided into groups, one that receives the novel form of vector control intervention, and comparison arms that often receive the old form of vector control tools or nothing. The key difficulty is that it is impossible to address the question to whether vector control would produce a smaller reduction in malaria if the vector mosquitoes are resistant than it would have done if they were susceptible, using RCT methods. This is simply because resistance is not an easy factor that can be allocated randomly to some communities and not to others. The distribution of resistance is patchy and its severity seems to differ from one location (village) to another. Moreover there may be more resistance or survival trend of mosquitoes in some villages than others because of variations in the quality of vector control operations, or in mosquito behavior [228, 229]. This is important to mention, because many health scientists regard evidence from randomized-controlled studies as the only reliable basis for decision-making in public health.

The first RCT that investigated the impact of pyrethroid-resistance on LLIN efficacy was conducted in the Korhogo area in the north of Côte d'Ivoire. The trial encompassed multiple villages where the 1014F *kdr*allele frequency was >90% [28] and malaria was endemic. The regular use of conventionally lambdacyhalothrin-treated nets had a significant impact on the entomological inoculation rate (55% reduction) and on malaria incidence in children < 5 (56% reduction of clinical attacks) compared to a control group having no nets [230]. This was the first clear-cut evidence of ITNs continuing to provide effective personal protection against malaria in an area with a very high frequency of *kdr* in the vector population. However, as reported in Ranson et al. [6], the absence of a physical barrier in the control group may have overestimated the impact of pyrethroid treated nets against *kdr* mosquitoes in this study.

More recently, another RCT of LLINs and/or IRS was conducted in 28 villages in southern Benin, from 2007 to 2010 [231]. The objective of the study was to examine whether carbamate-IRS applied every 8 months, as practiced by the PMI programme in Benin provided additional benefit over LLINs (ie Permanet 2.0) in term of malaria prevention and management of pyrethroid resistance in malaria vectors. Results showed that combination of LLINs and IRS did not reduce malaria transmission and morbidity compared to LLIN alone in an area of pyrethroid resistance [124]. Significant increase of 1014F *kdr* frequency was observed in the reference and treated arms only 18 months post intervention hence indicating that LLIN and IRS failed to reduce the spread of the 1014F allele in malaria vectors. The authors suggested that the increase in pyrethroid resistance might have accounted for the reduction of LLIN efficacy at a community level. Clearly, further investigation is needed to assess whether pyrethroid-resistance can reduce efficiency of LLINs and IRS for malaria prevention in Africa.

Given the many obstacles for evaluating the epidemiological impact of resistance, other alternative methods to measure operational impact has been to measure proxy entomological outcomes, such as the relative mortality and feeding success of resistant and susceptible vectors in experimental huts [232, 233]. Although such results can be remarkably clear, and definitively linked to resistance, experimental hut methods have their own limitations owing to the controlled hut structures that differ in many ways to normal houses in rural African context.

An early experimental hut trial of ITNs was conducted in the western African country of Benin. In southern Benin (Ladji), pyrethroid resistance has evolved in the M form of *An. gambiae* mosquitoes that appear to combine the knockdown resistance (*kdr*) gene with oxidase mechanisms [127, 234]. In Ladji, carrier mosquitoes of this resistance were not controlled by pyrethroid treatments in experimental hut trials of ITNs or the leading brands of LLINs, PermaNet 2.0 (Vestergaard Frandsen SA, Aarhus, Denmark) and Olyset (Sumitomo Chemicals, Osaka, Japan) [235]), compared to Malanville in the north where the vector was largely susceptible to pyrethroids [127]. Further household randomized trial conducted in northern susceptible and southern resistance areas demonstrated that lambdacyalothrin-ITNs (regardless the physical condition) lose their capacity to confer personal protection against pyrethroid-resistant A*n. gambiae* [236].

One of the problems associated with many of these studies is that, due to the lack of molecular markers for alternative resistance mechanisms (i.e. metabolic or even cuticular and behavioral), the frequency of*kdr* alleles is frequently used as a proxy for resistance. It has been recently demonstrated in Southern Benin that *kdr* by itself in *An. gambiae* does not seem to bear more malaria parasites than in a susceptible [237] but this conception can be misleading when metabolic or other resistance mechanisms are predominant or combine with *kdr* to confer resistance. There is an urgent need for properly controlled large-scale trials to assess the impact of pyrethroid resistance on IRS and ITNs in Africa but also in different regions affected by malaria (e.g. Asia and Latin America). Such studies should use both entomological and epidemiological indices and should be conducted in areas where alternative resistance mechanisms are known to be responsible for pyrethroid resistance. Furthermore, these studies must consider the possibility of behavioural resistance as recently suggested in Benin [238] and Tanzania [111] and monitor for changes in key traits such as location of resting and feeding which may impact on the efficacy of current insecticide based interventions.

RESISTANCE MANAGEMENT STRATEGIES

As a general statement, the use of insecticides does not create resistance by itself but select small proportion of individuals having a genetic mutation that allow them to resist and survive the effects of the insecticide. If this advantage is maintained by constant use of the same insecticide, the resistant insects will reproduce and the genetic changes that confer resistance will be transferred to offspring so that they become more prevalent within the population (figure 3). This selection process will take longer time to occur if the gene conferring resistance is rare or present at a low prevalence. Resistance should not be confused with "induction" that can occur after sub-lethal (or low dose) exposure to any insecticide and/or xenobiotic and is not passed on to offspring [239].

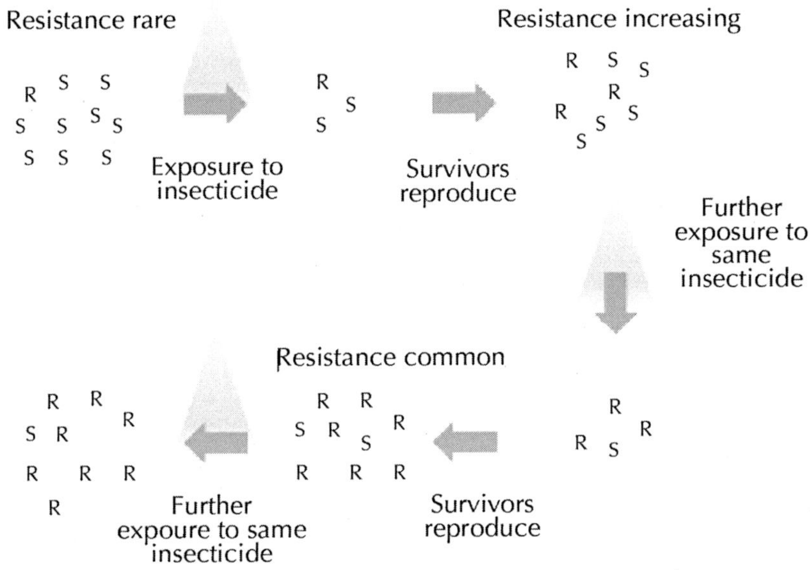

Figure 3: Possible scenario for resistance development in a mosquito population (source; [240])

Main Factors Influencing Resistance Development

The evolution of insecticide resistance is complex and depends on several genetic, biological and operational factors [240-242]. The biological factors relate by the life cycle of the insect (e.g. rate of reproduction, number of generation/offspring, rate of migration and isolation, etc), while the genetic factors include the intrinsic characteristics of the resistant genes (e.g. mono *versus* polygenic resistance, dominance, fitness cost and gene interaction). Operational factors concern the treatment itself including the method and frequency of application, dosage and residual activity of the insecticides as well as insecticide coverage.

Biological Factors

Rate of Reproduction

Insect species that have a short life cycle and high rates of reproduction are likely to develop resistance more rapidly than species that have a lower rate of reproduction, as any resistance genes can rapidly spread throughout the population. Because mosquitoes can produce high number of offspring (i.e. females can lay several hundred eggs during their reproductive life) they are much likely to develop resistance to insecticides than other species.

Population Migration / Isolation

With mosquitoes, the goal is to eliminate all or the majority of the population, however the greater the selection pressure that is put on a population, the faster susceptibility may be lost. Immigration of individuals possessing susceptible genes from untreated areas can beneficially dilute and compete with the resistance genes in the overall population. An early step in a malaria vector control programme should therefore be to estimate the susceptibility status of vector populations (see section 5 for details) and estimate potential immigration of untreated insects. This can be achieved by using genetic markers to estimate the gene flow (migrants) and genetic structure between populations. For example, an isolated area (e.g. island) where the entire area is treated would have a higher risk of developing resistance as few "susceptible" genotypes would join the treated population. The risk of insecticide resistance developing should be considered when planning a resistance management strategy. Awareness of and coordination with neighboring vector control programmes and agricultural activities should be encouraged, so that the regional and potential "side effect" on the target population is considered.

Genetics Factor

Dominance

Resistance genes can range from dominant through semi-dominant to recessive. If dominant or semi-dominant, only one parent needs to possess the characteristic to be fully or partially expressed in the offspring. If recessive, both parents must possess the trait. Fortunately, most resistance mechanisms (e.g.*kdr*) are controlled by recessive or semi-recessive genes, which slows their spread within the population at early stage of resistance development when most individuals are present at heterozygous state. In contrast, when the resistance is genetically dominant, e.g. the *Ace.1*^{R} gene conferring cross-resistance to carbamates and OPs [243], it can rapidly become established within the population and will be difficult to manage. Fortunately, strong genetic cost is often associated with dominant resistant gene that can compensate the effect of the dominance and slow down the increase of resistance gene frequency in natural populations [244].

Gene Interactions

Epistasis is the non-additive interaction (synergistic *versus* antagonistic) between different loci which contribute to a phenotype [245]. Epistasis between independent loci conferring insecticide resistance is important to investigate as this phenomenon can shape the rate at which resistance evolves and can dictate the level of resistance in the field. Epistasis can be measured in laboratory studies on susceptible and resistant colonies, but without these data, it is generally impossible to predict whether or not it will occur when two genes are being evaluated. Studies of the interactions between resistance loci have been most commonly conducted in house flies [246-248]. Generally, a greater than additive interaction was observed between two loci that were both homozygous resistant, whereas additively (i.e. lack of epistasis) occurs between two loci that were both heterozygous. In mosquitoes, Harstone and colleagues [249] showed multiplicative interactions between *kdr* and P450 detoxification in *Culex pipiens quinquefasciatus* whether the resistance alleles were homozygous or heterozyges. For example, resistance ratio 50 (i.e. LC50 resistant strain / LC50 susceptible strain) to permethrin in the double homozygote mosquitoes (RR50 of 30,000) was much higher than that expected (RR50 of 1,400) by simple additive effect of the two loci. Overall, interactions between independent resistant genes are complex. It is therefore important to better understand the interactions between resistant loci as well as to address how the fitness costs/benefits of the mechanisms can manipulate the observed interactions.

Fitness Cost

Populations of insects that have never been exposed to insecticides are usually fully susceptible, and resistance genes within those populations are very rare. This usually occurs through a "fitness cost", which means that insects sharing the resistance allele lack some other attribute or "quality" such that it gives an advantage to the susceptible insects in an insecticide free environment [250]. For example, resistant insects may have lower mating success, be more susceptible to natural enemies [251], or more prone to mortality during over-wintering [252]. Increased production of metabolic enzymes generally shows lower associated fitness cost than those associated with alterations in the structural genes most probably because the primary function of the enzyme is not disrupted [253]. There is good laboratory and field evidence to suggest that the deficit of insecticide selection pressure, in most cases, selects for susceptible genotypes. For example, the absence of homozygote's resistant genotypes in *An. gambiae*populations in West Africa is most probably due to the strong genetic cost associated with the carbamate-resistance allele *ace.1*^{R} (G119S) [254]. In addition, once resistance in the field has been selected it often rapidly reverts once the insecticide treatment regime is changed. A good example of this occurred in *An. arabiensis* in Sudan, where malathion specific insecticide resistance was selected in the early 1980s through antimalarial house spraying. The development of resistance prompted a switch of insecticide treatment to fenitrothion and the malathion resistance rapidly reverted in the following years. However, reversion rates are variable and may be very slow, particularly when an insecticide has been used for many years. For example, DDT was used extensively for malaria control over a 20 year period up to the 1960s in Sri Lanka to control *An. culicifacies s.l.* and *An. subpictus*. DDT was replaced by malathion in Sri Lanka in the early 1970s when a total and effective ban on DDT use was implemented. Subsequent regular monitoring has shown that DDT resistance has reverted very slowly towards susceptibility; around 80% of the adult mosquito population was resistant in the 1970s compared to about 50% in the 1990s. The same is true with the R*dl* gene that was maintained in field mosquito populations despite the abandon of cyclodiene for mosquito control for more than 30 years [33]. Rate of reversion is an important parameter to consider before implementing any resistance management strategy in the field.

Operational Factors

In practice, only operational factors such as the insecticide(s) used, the area of coverage (for example for IRS or LLIN), and the timing, rate, and method of application can be manipulated directly to reduce the selection pressure for resistance. Operational factors influence selection by determining the overall fraction of a population exposed (larvae/adults) to a selecting agent and the degree of contact and pick-up of toxicant by exposed pests at what has been termed the "interface between insects and insecticides" [242]. At both stages, operational and intrinsic factors interact in complex ways to establish the net effect of a control treatment on both genetic composition and total population size. Management of resistance therefore entails resolving these interactions to anticipate with some confidence both the suppressive and selective effects of potential control strategies.

Frequency of Application, Dosage and Persistence of Effect

How often an insecticide is used is one of the most important factors that influence resistance development [240]. With each use, an advantage is given to the resistant insects within a population. The rate of increase of resistance on any population will generally be faster in the presence of a lower fitness cost and high reproductive and short life cycles producing several generations per season. The length of time that an insecticide remains effective, also called its persistence, is dependent upon the physical chemistry of the insecticide, the type of formulation, the application rate and the substrate. Products which provide a persistent effect provide continual selection pressure in a similar manner to multiple treatments. For example, a space spray will persist for a very short time and will select only against a single generation of mosquitoes. In contrast, a residual wall application (IRS) or Insecticide Treated nets treatment (especially Long Lasting Nets) will persist for months or years providing a selection pressure against many generations of the same insect. For example, repeated application of DDT for indoor residual spraying has contributed to increase the number of DDT-resistant malaria vector species in various geographical settings [255]. Several studies showed however that the use of insecticides in agriculture play a key role in the selection of resistance in mosquitoes

[256, 257]. Indeed, most insecticides used in agriculture are of the same chemical classes and have the same targets and modes of action as those used in public health programme. In practice, VC programmes cannot influence the choice of the pesticide used for crop protection and the only thing that can be done is to appropriately select the most judicious insecticide for mosquito control. However, there is more published evidence that public health insecticides can contribute to select for pyrethroid resistance alleles (see section 7 for details). It is obvious that we can expect enhanced selection pressure on resistance genes through the scaling up of LLIN and/or IRS for malaria elimination.

Choice of the Insecticide

The speed at which an insecticide effectively kills an insect can also influence the evolution of resistance. All current insecticides approved for ITNs or IRS kill extremely rapidly after contact. While fast-acting conventional insecticides can produce even more effective initial control, they impose enormous selection for resistance by killing young female adults. The consequence is that spectacular initial mosquito control can last as little as a few years, thus providing very poor medium- to long-term disease control [258]. Some authors recently suggest that Late Acting insecticides (e.g. entomofungus) may be a more tactical strategy to manage resistance if female mosquitoes are killed after 2 or more gonotrophic cycles [259]. Indeed, the less the insecticide impact on mosquito fitness, the less the strength of selection, especially if the resistance allele is associated with a strong genetic cost. In theory, it would be possible to create an insecticide that would provide effective malaria control yet never be undermined by the evolution of resistant mosquitoes. However, further studies are required as "proof of principle" i.e. to demonstrate that this strategy can be effective for vector management and malaria prevention in a real setting.

Resistance Management — Strategies and Tactics

Historically, the practice of using an insecticide until resistance becomes a limiting factor has rapidly eroded the number of suitable/

available insecticides for vector control. Rotations, mosaics, and mixtures have all been proposed as resistance management tools [260, 261] but there are very few "success stories" in public health. Numerous mathematical models have been produced to estimate how these tools could be optimally used [262-264] but these models have rarely been tested under field conditions due to the practical difficulties in estimating changes in resistance gene frequencies (especially for metabolic resistance) in large samples of insects [220]. With the advent of different molecular techniques for resistance-gene frequency estimation, field trials of resistance management strategies have now become more feasible.

Approaches to Resistance Management

Ideally insecticide resistance management should be undertaken using insecticide based approaches in conjunction with other non-insecticidal vector control methods, i.e. as part of Integrated Vector Management (IVM[4] -) [265]. The insecticides used have to be safe to humans and comply with WHO specifications. In practice, most of IVM programmes work well in experimental trials but become challenging when programmes are scaling up into long-term (operational) control. Operationally, the simplest form of resistance management is likely to be "insecticide" based, and this could take several forms.

Rotations

Rotational strategies are based on the rotation over time of two or preferably more insecticide classes with different modes of action. This approach assumes that if resistance to each insecticide is rare, then multiple resistances will be extremely rare [266]. Rotation allows any resistance developed to the first insecticide to decline over time when the second insecticide class is introduced. As for other strategy, rotations are particularly effective if the resistance gene has an associated fitness cost. The timeframe for rotation needs to be sufficiently short to prevent significant levels of resistance to develop to any one rotation partner. Rotations have been successful in many applications in agriculture and are considered to be effective in slowing the evolution of resistance (see [240] for details). The most striking example of "success story" using this strategy was within the framework of the Onchocerciasis Control Programme (OCP) carried in West Africa 40 years ago. Indeed, weekly application of unrelated larvicides in rivers was successful to kill the larvae of the blackfly vector and mitigate the spread of temephos resistance over the 17 years of its implementation [267]. However, the rotation was introduced at early stage of the OCP, as soon as the operators faced temephos resistance problems in pilot localities. As for all IRM strategies, the status of resistance of the insecticide used in the rotation must be known when implementing rotations and the chemicals used should not present any (known) cross-resistance. For LLIN, it is difficult to implement this method knowing that only pyrethroids are recommended so far by WHO for the impregnation [268]. For IRS, the pragmatic approach would be to rotate insecticides annually. Indeed, changing insecticides more than once a year (which could be the case in areas where two spray rounds are conducted each year) is not recommended, because of procurement and other financial and logistical challenges (see [4]). Despite higher cost of implementing rotation than single spray (as available alternatives to pyrethroids —the carbamates organophosphates, insect growth regulators, pyroles —are currently more expensive), this is probably the price to pay to preserve the arsenal of cost-effective insecticides for malaria vector control.

Mosaics

Spatially separated applications of different compounds against the same insect constitute a "mosaic" approach to resistance management [240]. Fine scale mosaics can be achieved in malaria vector control programmes, for example, by using two insecticides in different dwellings within the same village. The aim of this strategy is to preserve susceptibility by spatial restriction of insecticides [4]. If such a fine scale mosaic is to be used, careful records of which insecticide was used in each house are essential. Larger scale mosaics have been shown to be effective for the management of pyrethroid resistance in *An. albimanus* in Mexico [22]. Indeed, pyrethroid resistance rose more rapidly in the areas under pyrethroid treatment alone than in the mosaic areas using OP, Pyrethroid and carbamate [240]. Whilst there are some practical difficulties implementing a mosaic in a vector control programme (eg spray with different insecticides, dosages, apparatus, etc), it may offer the advantages of a mixture strategy with lower insecticide inputs and hence cost. The scale at which a mosaic needs to be applied has not been clearly established. In South Africa for example, different insecticides have been used in different types of houses within the same community and this is considered by some to be a mosaic-like strategy [240]. Similarly, mosquito bed nets from panels treated with different insecticides achieve a similar mosaic effect to treating houses with different compounds but on a much finer scale. Industry has recently developed mosaic LLINs containing a pyrethroid insecticide and a synergist (Piperonyl butoxide or PBO an oxidase inhibitor) on the roof to increase efficacy against pyrethroid-resistant malaria vectors. Small scale field trial [235, 269] and mathematical exercises [270] suggested that mosaic LLIN may provide better insecticidal effect against resistant mosquitoes and enhanced community-level protection against malaria compared to "classical" LLIN in area of pyrethroid resistance. Clearly further operational research is required to establish the applicability and effectiveness of mosaics approaches for malaria control.

Mixtures

A mixture is defined by the simultaneous use of two or more insecticides of unrelated mode of action. If two insecticides A and B, with independent resistance mechanisms, are applied together in a mixture, and if resistance to A and resistance to B are both rare, then we expect doubly resistant insects to be extremely rare, and almost all insects resistant to A will be killed by B, and vice versa [266]. This system of "redundant killing" means that resistance to the two insecticides will evolve much more slowly than if either had been used on its own [271]. This approach may be not successful if resistance to one of the components used is already present at a detectable level and/or if linkage disequilibrium is present in the targeted population [4]. Unlike rotations, the effectiveness of mixtures is not directly related to the degree of fitness cost. Rather the mixture aims to overpower resistance instead of preserving susceptibility. However, for mixtures to work well in practice both insecticides need to be used at their full application rate in order that the efficacy and persistence of the two insecticides would be broadly similar (same decay rate). Further, theoretical models suggest that mixtures might delay resistance longer than rotations or broad mosaics [271, 272]. However, mixtures of products were rarely adopted in malaria vector control programmes on grounds of cost, logistics, and safety issue and because of the limited number of recommended compounds available for both IRS and LLIN. It is not yet clear however how much the addition of a second active ingredient will add to the total cost of manufacturing since the cost of additional insecticide can greatly vary according to the strategy, ie. cost for LLINs would be much lower than that for IRS. For LLIN, previous laboratory and field trials showed interesting prospects for reducing mosquito survival and biting rates with the use of insecticide mixtures applied on mosquito nets against*kdr*-resistant *An. gambiae* in Africa [273, 274]. Other chemicals, such as insect growth regulators (IGR), represent also promising alternative to be included in mixture formulations as they may impact on mosquito longevity, fertility and fecundity [275, 276]. With the development of next-generation of LLIN, combined use of non-pyrethroids and pyrethroids on bed nets is technically achievable and has the potential to provide better control of malaria and prevent further development of pyrethroid-resistance in malaria vectors. Risk assessment and acceptability of such new tools should be however carefully investigated before any trial being implemented at operational level.

Combinations

In this context, combinations expose the vector population to two vector control tools, such that a mosquito that survives contact with one (e.g. LLIN) is exposed to the other one (e.g. IRS), or vice versa. In practice, exposure to two insecticides is not guaranteed but there is some evidence to indicate that this is likely [277]. The effectiveness of combinations in IRM does not depend on the ability to reduce the level of resistance, but on the ability to kill the vector despite the existence of resistance, through the use of another insecticide or intervention, which compensates for resistance [231]. As for other strategy, the combination should not contain insecticides with same mode of action (e.g. avoid pyrethroids for both IRS and LLINs), as this would increase selection pressure rather than reducing it. As combinations require doubling of interventions, cost would be significantly higher than rotations and mosaics. This might nevertheless be warranted in some circumstances, for example where malaria transmission is very high and/or where targeted IRS can help overcome identified resistance to pyrethroids in areas with high LLIN coverage. In practice, combinations would be more easily implemented in countries having sufficient human and financial resources allocated to public health programmes. So far, a small number of observational studies [278-280] and mathematical modeling exercises [263, 264] suggest that VC combination has an added benefit for reduction of the risk of infection because the people not protected by one of the interventions are protected by the other. A recent cluster randomized controlled trial carried out in Benin showed however that neither clinical malaria in children younger than 6 years nor transmission intensity differ between LLIN and carbamate-IRS or Carbamate Treated Plastic Sheeting and the reference group (LLIN alone) and the insecticide combinations did not slow down the evolution of the*kdr* allele in *An gambiae s.s.* compared with LLIN [231]. It was concluded from this study that IRS should be timely implemented (i.e. using appropriate insecticide, dosage and time interval) to ensure optimum efficacy of the IRS intervention over LLIN. Clearly, cost-effectiveness of combined vector control interventions need to be carefully considered to ensure that increased efforts and cost dedicated to combinations effectively contribute to better control and management of pyrethroid-resistant malaria vectors.

CONCLUSIONS

Insecticide resistance develops in an insect population when individuals carrying genes that allow them to survive exposure to the insecticide pass these genes on. Thus, any activities that control the individuals with the resistance trait will delay the spread of the resistance genes in the population. IRM should then be seen in the context of IVM and should therefore also include activities such as habitat management, community education, and/or larval source management (e.g. biological control). In order to successfully develop and implement any resistance management strategies based on rotations, mosaics, mixtures or combinations, knowledge of the mode of action, chemical properties, and residual life of the available insecticide products is essential. Although insecticides with novel modes of action have recently been introduced in public health (neonicotinoids, pyroles, oxadiazin, etc) few of them appear to have the optimum biological and/or physical properties required for residual wall spray and/or mosquito net. Unfortunately, the exorbitant costs associated with developing and registering new insecticides (see [281] for details) mean that products appear in the more profitable agricultural markets before consideration is given to their public health potential. We have then no other option than to make an appropriate and judicious use of the current insecticides if we want to avoid any disillusion with pyrethroids as we faced before with DDT or dieldrin. The philosopher George Santayana said *"those who cannot remember the past are condemned to repeat it."* Hope it's not too late for malaria vector control.

REFERENCES

1. World Health Organization: WHO malaria report : 2011. In: *WHO Global Malaria Programme*. Edited by Data WLC-i-P. Geneva: WHO; 2011: 259.
2. Yassine H, Osta MA: *Anopheles gambiae* innate immunity. *Cell Microbiol*, 12(1):1-9.
3. Davidson G: Insecticide resistance in *Anopheles sundaicus*. *Nature* 1957, 180(4598):1333-1335.
4. World Health Organization: Global Plan for Insecticide Resistance Management in Malaria Vectors (GPIRM). In: *WHO/HTM/GMP/20125*. Edited by Organization WH. Geneva, Switzerland: World Health Organization; 2012: 130.
5. Hemingway J, Ranson H: Insecticide resistance in insect vectors of human disease. *Annu Rev Entomol* 2000, 45:371-391.
6. Ranson H, N'Guessan R, Lines J, Moiroux N, Nkuni Z, Corbel V: Pyrethroid resistance in African anopheline mosquitoes: what are the implications for malaria control? *Trends Parasitol* 2011, 27(2):91-98.
7. Ridl FC, Bass C, Torrez M, Govender D, Ramdeen V, Yellot L, Edu AE, Schwabe C, Mohloai P, Maharaj R *et al*: A pre-intervention study of malaria vector abundance in Rio Muni, Equatorial Guinea: their role in malaria transmission and the incidence of insecticide resistance alleles. *Malar J*2008, 7:194.
8. Sharp BL, Ridl FC, Govender D, Kuklinski J, Kleinschmidt I: Malaria vector control by indoor residual insecticide spraying on the tropical island of Bioko, Equatorial Guinea. *Malar J* 2007, 6:52.
9. Mitchell SN, Stevenson BJ, Muller P, Wilding CS, Egyir-Yawson A, Field SG, Hemingway J, Paine MJ, Ranson H, Donnelly MJ: Identification and validation of a gene causing cross-resistance between insecticide classes in *Anopheles gambiae* from Ghana. *Proc Natl Acad Sci U S A* 2012, 109(16):6147-6152.

10. Stevenson BJ, Bibby J, Pignatelli P, Muangnoicharoen S, O'Neill PM, Lian LY, Muller P, Nikou D, Steven A, Hemingway J *et al*: Cytochrome P450 6M2 from the malaria vector *Anopheles gambiae* metabolizes pyrethroids: Sequential metabolism of deltamethrin revealed. *Insect Biochem Mol Biol* 2011, 41(7):492-502.
11. Brown AWA: Insecticide Resistance Mosquitoes: A Pragmatic Rewiew. *J Am Mosq Control Assoc* 1986, 2(2):123-140.
12. Metcalf RL: Insect resistance to insecticides. *Pesticide Science* 1989, 26(4):333-358.
13. WHO: Global Insecticide Use for Vector-Borne Disease Control-4th edition. *In Edited by World Health Organization Geneva: World Health Organization* 2009, WHO/HTM/NTD/WHOPES/GCDPP/2009.6.
14. Lapied B, Pennetier C, Apaire-Marchais V, Licznar P, Corbel V: Innovative applications for insect viruses: towards insecticide sensitization. *Trends Biotechnol* 2009, 27(4):190-198.
15. Livadas GA, Georgopoulos G: Development of resistance to DDT by *Anopheles sacharovi* in Greece. *Bull World Health Organ* 1953, 8(4):497-511.
16. Brown AWA, Haworth J, Zahar AR: Malaria eradication and control from a global standpoint. *Journal of Medical Entomology* 1976, 13(1):1-25.
17. Patel TB, Ramachandra Rao T, Halgeri AV, Deobhankar RB: A preliminary note on a probable case of dieldrin kesistance in *Anopheles culicifaciesin* Thana District, Bombay State. *Indian J Malariol* 1958, 12:367-370.
18. Brown AW, Pal R: Insecticide resistance in arthropods. *Public Health Pap* 1971, 38(0):1-491.
19. Brown AW, Haworth J, Zahar AR: Malaria eradication and control from a global standpoint. *J Med Entomol* 1976, 13(1):1-25.
20. Rajagopal R: Malathion resistance in *Anopheles culicifacies* in Gujarat. *Indian J Med Res* 1977, 66(1):27-28.
21. Rawlings P, Herath PR, Kelly S: *Anopheles culicifacies* (Diptera: Culicidae): DDT resistance in Sri Lanka prior to and after cessation of DDT spraying. *J Med Entomol* 1985, 22(4):361-365.

22. Penilla RP, Rodriguez AD, Hemingway J, Torres JL, Arredondo Jimenez JI, Rodriguez MH: Resistance management strategies in malaria vector mosquito control. Baseline data for a large-scale field trial against *Anopheles albimanus* in Mexico. *Medical and Veterinary Entomology* 1998, 12(3):217-233.
23. Van Bortel W, Trung HD, Thuan le K, Sochantha T, Socheat D, Sumrandee C, Baimai V, Keokenchanh K, Samlane P, Roelants P *et al*: The insecticide resistance status of malaria vectors in the Mekong region. *Malar J* 2008, 7:102.
24. Trung HD, Van Bortel W, Sochantha T, Keokenchanh K, Quang NT, Cong LD, Coosemans M: Malaria transmission and major malaria vectors in different geographical areas of Southeast Asia. *Trop Med Int Health* 2004, 9(2):230-237.
25. Chareonviriyaphap T, Aum-aung B, Ratanatham S: Current insecticide resistance patterns in mosquito vectors in Thailand. *Southeast Asian J Trop Med Public Health* 1999, 30(1):184-194.
26. Hamon J, Subra R, Sales S, Coz J: [Presence in the southwestern part of Upper Volta of a population of *Anopheles gambiae* "A" resistant to DDT].*Med Trop (Mars)* 1968, 28(4):521-528.
27. Carnevale P, Robert V: *Anopheles*: Biologie, transmission du *Plasmodium* et lutte antivectorielle. Bondy: IRD; 2009.
28. Chandre F, Darriet F, Manguin S, Brengues C, Carnevale P, Guillet P: Pyrethroid cross resistance spectrum among populations of *Anopheles gambiae* s.s. from Cote d'Ivoire. *J Am Mosq Control Assoc* 1999, 15(1):53-59.
29. Armstrong JA, Ramsdale CD, Ramakrishna V: Insecticide resistance in *Anopheles gambiae* Giles in Western Sokoto, Northern Nigeria. *Ann Trop Med Parasitol* 1958, 52(3):247-256.
30. Hamon J, Garrett-Jones C: [Resistance to insecticides in the major malaria vectors and its operational importance]. *Bull World Health Organ*1963, 28(1):1-24.
31. Wondji CS, Dabire RK, Tukur Z, Irving H, Djouaka R, Morgan JC: Identification and distribution of a GABA receptor mutation conferring dieldrin resistance in the malaria vector *Anopheles funestus* in Africa. *Insect Biochem Mol Biol* 2011, 41(7):484-491.
32. Mouchet J: Mini-Review : Agriculture and Vector Resistance. *Insect Sci Applic* 1988, 9(3):291-302.

33. Mahande AM, Dusfour I, Matias JR, Kweka EJ: Knockdown Resistance, rdl Alleles, and the Annual Entomological Inoculation Rate of Wild Mosquito Populations from Lower Moshi, Northern Tanzania. *J Glob Infect Dis* 2012, 4(2):114-119.

34. Brogdon WG, McAllister JC, Corwin AM, Cordon Rosales C: Oxidase-based DDT-pyrethroid cross-resistance in Guatemalan *Anopheles albimanus.Pesticide Biochemistry and Physiology* 1999, 64(2):101-111.

35. Fonseca-Gonzalez I, Quinones ML, McAllister J, Brogdon WG: Mixed-function oxidases and esterases associated with cross-resistance between DDT and lambda-cyhalothrin in *Anopheles darlingi* Root 1926 populations from Colombia. *Mem Inst Oswaldo Cruz* 2009, 104(1):18-26.

36. Raghavendra K, Verma V, Srivastava HC, Gunasekaran K, Sreehari U, Dash AP: Persistence of DDT, malathion & deltamethrin resistance in*Anopheles culicifacies* after their sequential withdrawal from indoor residual spraying in Surat district, India. *Indian J Med Res* 2010, 132:260-264.

37. Tiwari S, Ghosh SK, Ojha VP, Dash AP, Raghavendra K: Reduced susceptibility to selected synthetic pyrethroids in urban malaria vector *Anopheles stephensi*: a case study in Mangalore city, South India. *Malar J* 2010, 9:179.

38. Elissa N, Mouchet J, Riviere F, Meunier JY, Yao K: Resistance of *Anopheles gambiae* s.s. to pyrethroids in Cote d'Ivoire. *Ann Soc Belg Med Trop*1993, 73(4):291-294.

39. Hargreaves K, Koekemoer LL, Brooke BD, Hunt RH, Mthembu J, Coetzee M: *Anopheles funestus* resistant to pyrethroid insecticides in South Africa. *Medical and Veterinary Entomology* 2000, 14(2):181-189.

40. Chareonviriyaphap T, Rongnoparut P, Juntarumporn P: Selection for pyrethroid resistance in a colony of *Anopheles minimus* species A, a malaria vector in Thailand. *J Vector Ecol* 2002, 27(2):222-229.

41. Ayad H, Georghiou GP: Resistance to organophosphates and carbamates in *Anopheles albimanus* based on reduced sensitivity of acetylcholinesterase. *Journal of Economic Entomology* 1975, 68(3):295-297.

42. Hemingway J: Genetics of organophosphate and carbamate resistance in *Anopheles atroparvus* (Diptera: Culicidae). *J Econ Entomol* 1982, 75(6):1055-1058.
43. Elissa N, Mouchet J, Riviere F, Meunier JY, Yao K: [Susceptibility of *Anopheles gambiae* to insecticides in the Ivory Coast]. *Sante* 1994, 4(2):95-99.
44. N'Guessan R, Darriet F, Guillet P, Carnevale P, Traore-Lamizana M, Corbel V, Koffi AA, Chandre F: Resistance to carbosulfan in *Anopheles gambiae* from Ivory Coast, based on reduced sensitivity of acetylcholinesterase. *Med Vet Entomol* 2003, 17(1):19-25.
45. Dabire KR, Diabate A, Namontougou M, Djogbenou L, Kengne P, Simard F, Bass C, Baldet T: Distribution of insensitive acetylcholinesterase (ace-1R) in *Anopheles gambiae* s.l. populations from Burkina Faso (West Africa). *Trop Med Int Health* 2009, 14(4):396-403.
46. Djogbenou L, Dabire R, Diabate A, Kengne P, Akogbeto M, Hougard JM, Chandre F: Identification and geographic distribution of the ACE-1R mutation in the malaria vector *Anopheles gambiae* in south-western Burkina Faso, West Africa. *Am J Trop Med Hyg* 2008, 78(2):298-302.
47. Djogbenou L, Pasteur N, Akogbeto M, Weill M, Chandre F: Insecticide resistance in the *Anopheles gambiae* complex in Benin: a nationwide survey.*Med Vet Entomol* 2011, 25(3):256-267.
48. Oduola AO, Idowu ET, Oyebola MK, Adeogun AO, Olojede JB, Otubanjo OA, Awolola TS: Evidence of carbamate resistance in urban populations of*Anopheles gambiae* s.s. mosquitoes resistant to DDT and deltamethrin insecticides in Lagos, South-Western Nigeria. *Parasit Vectors* 2012, 5:116.
49. USAID: The President's Malaria Initiative: fourth annual report. In. Washigton: U.S. Agency for International Development; 2010.
50. Nkya TE, Akhouayri I, Kisinza W, David JP: Impact of environment on mosquito response to pyrethroid insecticides: Facts, evidences and prospects. *Insect Biochem Mol Biol* 2012.
51. Despres L, David JP, Gallet C: The evolutionary ecology of insect resistance to plant chemicals. *Trends Ecol Evol* 2007, 22(6):298-307.

52. Hemingway J, Hawkes NJ, McCarroll L, Ranson H: The molecular basis of insecticide resistance in mosquitoes. *Insect Biochem Mol Biol* 2004, 34(7):653-665.

53. Guillemaud T, Makate N, Raymond M, Hirst B, Callaghan A: Esterase gene amplification in *Culex pipiens*. *Insect Molecular Biology* 1997, 6(4):319-327.

54. Hawkes NJ, Hemingway J: Analysis of the promoters for the beta-esterase genes associated with insecticide resistance in the mosquito *Culex quinquefasciatus*. *Biochim Biophys Acta* 2002, 1574(1):51-62.

55. Vaughan A, Hawkes N, Hemingway J: Co-amplification explains linkage disequilibrium of two mosquito esterase genes in insecticide-resistant*Culex quinquefasciatus*. *Biochem J* 1997, 325 (Pt 2):359-365.

56. Herath PR, Miles SJ, Davidson G: Fenitrothion (OMS 43) resistance in the taxon *Anopheles culicifacies* Giles. *J Trop Med Hyg* 1981, 84(2):87-88.

57. Hemingway J: The genetics of malathion resistance in *Anopheles stephensi* from Pakistan. *Trans R Soc Trop Med Hyg* 1983, 77(1):106-108.

58. Vontas J, Blass C, Koutsos AC, David JP, Kafatos FC, Louis C, Hemingway J, Christophides GK, Ranson H: Gene expression in insecticide resistant and susceptible *Anopheles gambiae* strains constitutively or after insecticide exposure. *Insect Mol Biol* 2005, 14(5):509-521.

59. Somwang P, Yanola J, Suwan W, Walton C, Lumjuan N, Prapanthadara LA, Somboon P: Enzymes-based resistant mechanism in pyrethroid resistant and susceptible *Aedes aegypti* strains from northern Thailand. *Parasitol Res* 2011, 109(3):531-537.

60. Feyereisen R, Lawrence IG, Kostas I, Sarjeet SG: Insect Cytochrome P450. In: *Comprehensive Molecular Insect Science*. Amsterdam: Elsevier; 2005: 1-77.

61. Holt RA, Subramanian GM, Halpern A, Sutton GG, Charlab R, Nusskern DR, Wincker P, Clark AG, Ribeiro JM, Wides R *et al*: The genome sequence of the malaria mosquito *Anopheles gambiae*. *Science* 2002, 298(5591):129-149.

62. Ranson H, Nikou D, Hutchinson M, Wang X, Roth CW, Hemingway J, Collins FH: Molecular analysis of multiple cytochrome P450 genes from the malaria vector, *Anopheles gambiae*. *Insect Mol Biol* 2002, 11(5):409-418.

63. Nikou D, Ranson H, Hemingway J: An adult-specific CYP6 P450 gene is overexpressed in a pyrethroid-resistant strain of the malaria vector,*Anopheles gambiae*. *Gene* 2003, 318:91-102.

64. David JP, Strode C, Vontas J, Nikou D, Vaughan A, Pignatelli PM, Louis C, Hemingway J, Ranson H: The *Anopheles gambiae* detoxification chip: a highly specific microarray to study metabolic-based insecticide resistance in malaria vectors. *Proc Natl Acad Sci U S A* 2005, 102(11):4080-4084.

65. Irving H, Riveron JM, Ibrahim SS, Lobo NF, Wondji CS: Positional cloning of rp2 QTL associates the P450 genes CYP6Z1, CYP6Z3 and CYP6M7 with pyrethroid resistance in the malaria vector *Anopheles funestus*. *Heredity (Edinb)* 2012, 109(6):383-392.

66. Chiu TL, Wen Z, Rupasinghe SG, Schuler MA: Comparative molecular modeling of *Anopheles gambiae* CYP6Z1, a mosquito P450 capable of metabolizing DDT. *Proc Natl Acad Sci U S A* 2008, 105(26):8855-8860.

67. McLaughlin LA, Niazi U, Bibby J, David JP, Vontas J, Hemingway J, Ranson H, Sutcliffe MJ, Paine MJ: Characterization of inhibitors and substrates of *Anopheles gambiae* CYP6Z2. *Insect Mol Biol* 2008, 17(2):125-135.

68. Muller P, Chouaibou M, Pignatelli P, Etang J, Walker ED, Donnelly MJ, Simard F, Ranson H: Pyrethroid tolerance is associated with elevated expression of antioxidants and agricultural practice in *Anopheles arabiensis* sampled from an area of cotton fields in Northern Cameroon. *Mol Ecol*2008, 17(4):1145-1155.

69. Djouaka RF, Bakare AA, Coulibaly ON, Akogbeto MC, Ranson H, Hemingway J, Strode C: Expression of the cytochrome P450s, CYP6P3 and CYP6M2 are significantly elevated in multiple pyrethroid resistant populations of *Anopheles gambiae* s.s. from Southern Benin and Nigeria. *BMC Genomics* 2008, 9:538.

70. Muller P, Donnelly MJ, Ranson H: Transcription profiling of a recently colonised pyrethroid resistant *Anopheles gambiae* strain from Ghana. *BMC Genomics* 2007, 8:36.

71. Muller P, Warr E, Stevenson BJ, Pignatelli PM, Morgan JC, Steven A, Yawson AE, Mitchell SN, Ranson H, Hemingway J *et al*: Field-caught permethrin-resistant *Anopheles gambiae* overexpress CYP6P3, a P450 that metabolises pyrethroids. *PLoS Genet* 2008, 4(11):e1000286.

72. Wondji CS, Irving H, Morgan J, Lobo NF, Collins FH, Hunt RH, Coetzee M, Hemingway J, Ranson H: Two duplicated P450 genes are associated with pyrethroid resistance in *Anopheles funestus*, a major malaria vector. *Genome Res* 2009, 19(3):452-459.

73. Wondji CS, Morgan J, Coetzee M, Hunt RH, Steen K, Black WCt, Hemingway J, Ranson H: Mapping a quantitative trait locus (QTL) conferring pyrethroid resistance in the African malaria vector *Anopheles funestus*. *BMC Genomics* 2007, 8:34.

74. Riveron JM, Irving H, Ndula M, Barnes KG, Ibrahim SS, Paine MJ, Wondji CS: Directionally selected cytochrome P450 alleles are driving the spread of pyrethroid resistance in the major malaria vector *Anopheles funestus*. *Proc Natl Acad Sci U S A* 2012, 110(1):252-257.

75. Rongnoparut P, Boonsuepsakul S, Chareonviriyaphap T, Thanomsing N: Cloning of cytochrome P450, CYP6P5, and CYP6AA2 from *Anopheles minimus* resistant to deltamethrin. *J Vector Ecol* 2003, 28(2):150-158.

76. Duangkaew P, Kaewpa D, Rongnoparut P: Protective efficacy of *Anopheles minimus* CYP6P7 and CYP6AA3 against cytotoxicity of pyrethroid insecticides in *Spodoptera frugiperda* (Sf9) insect cells. *Trop Biomed* 2011, 28(2):293-301.

77. Duangkaew P, Pethuan S, Kaewpa D, Boonsuepsakul S, Sarapusit S, Rongnoparut P: Characterization of mosquito CYP6P7 and CYP6AA3: differences in substrate preference and kinetic properties. *Arch Insect Biochem Physiol* 2011, 76(4):236-248.

78. Ranson H, Cornel AJ, Fournier D, Vaughan A, Collins FH, Hemingway J: Cloning and Localization of a Glutathione S-transferase Class I Gene from*Anopheles gambiae*. *J Biol Chem* 1997, 272(9):5464-5967.

79. Ranson H, Hemingway J: Mosquito glutathione transferases. *Methods Enzymol* 2005, 401:226-241.

80. Fournier D, Bride JM, Poirie M, Berge JB, Plapp FW, Jr.: Insect

glutathione S-transferases. Biochemical characteristics of the major forms from houseflies susceptible and resistant to insecticides. *Journal of Biological Chemistry* 1992, 267(3):1840-1845.

81. Ranson H, Rossiter L, Ortelli F, Jensen B, Wang XL, Roth CW, Collins FH, Hemingway J: Identification of a novel class of insect glutathione S-transferases involved in resistance to DDT in the malaria vector *Anopheles gambiae*. *Biochemical Journal* 2001, 359 Part 2:295-304.
82. Gunasekaran K, Muthukumaravel S, Sahu SS, Vijayakumar T, Jambulingam P: Glutathione S transferase activity in Indian vectors of malaria: A defense mechanism against DDT. *J Med Entomol* 2011, 48(3):561-569.
83. Prapanthadara LA, Ketterman AJ: Qualitative and quantitative changes in glutathione S-transferases in the mosquito *Anopheles gambiae* confer DDT-resistance. *Biochem Soc Trans* 1993, 21 (Pt 3)(3):304S.
84. Ranson H, Jensen B, Wang X, Prapanthadara L, Hemingway J, Collins FH: Genetic mapping of two loci affecting DDT resistance in the malaria vector *Anopheles gambiae*. *Insect Mol Biol* 2000, 9(5):499-507.
85. Vontas JG, Small GJ, Hemingway J: Glutathione S-transferases as antioxidant defence agents confer pyrethroid resistance in *Nilaparvata lugens*.*Biochem J* 2001, 357(Pt 1):65-72.
86. Kostaropoulos I, Papadopoulos AI, Metaxakis A, Boukouvala E, Papadopoulou-Mourkidou E: Glutathione S-transferase in the defence against pyrethroids in insects. *Insect Biochem Mol Biol* 2001, 31(4-5):313-319.
87. Fournier D: Mutations of acetylcholinesterase which confer insecticide resistance in insect populations. *Chem Biol Interact* 2005.
88. Weill M, Lutfalla G, Mogensen K, Chandre F, Berthomieu A, Berticat C, Pasteur N, Philips A, Fort P, Raymond M: Comparative genomics: Insecticide resistance in mosquito vectors. *Nature* 2003, 423(6936):136-137.

89. Alout H, Labbe P, Berthomieu A, Pasteur N, Weill M: Multiple duplications of the rare ace-1 mutation F290V in *Culex pipiens* natural populations.*Insect Biochem Mol Biol* 2009.

90. Weill M, Malcolm C, Chandre F, Mogensen K, Berthomieu A, Marquine M, Raymond M: The unique mutation in ace-1 giving high insecticide resistance is easily detectable in mosquito vectors. *Insect Mol Biol* 2004, 13(1):1-7.

91. Djogbenou L, Chandre F, Berthomieu A, Dabire R, Koffi A, Alout H, Weill M: Evidence of introgression of the ace-1(R) mutation and of the ace-1 duplication in West African *Anopheles gambiae* s. s. *PLoS ONE* 2008, 3(5):e2172.

92. Ffrench-Constant RH: The molecular and population genetics of cyclodiene insecticide resistance. *Insect Biochem Mol Biol* 1994, 24(4):335-345.

93. Andreasen MH, Ffrench-Constant RH: In situ hybridization to the Rdl locus on polytene chromosome 3L of *Anopheles stephensi*. *Med Vet Entomol*2002, 16(4):452-455.

94. Du W, Awolola TS, Howell P, Koekemoer LL, Brooke BD, Benedict MQ, Coetzee M, Zheng L: Independent mutations in the Rdl locus confer dieldrin resistance to *Anopheles gambiae* and *An. arabiensis*. *Insect Mol Biol* 2005, 14(2):179-183.

95. Davies TG, Field LM, Usherwood PN, Williamson MS: DDT, pyrethrins, pyrethroids and insect sodium channels. *IUBMB Life* 2007, 59(3):151-162.

96. Donnelly MJ, Corbel V, Weetman D, Wilding CS, Williamson MS, Black WCt: Does *kdr* genotype predict insecticide-resistance phenotype in mosquitoes? *Trends Parasitol* 2009, 25(5):213-219.

97. Martinez Torres D, Chandre F, Williamson MS, Darriet F, Berge JB, Devonshire AL, Guillet P, Pasteur N, Pauron D: Molecular characterization of pyrethroid knockdown resistance (*kdr*) in the major malaria vector *Anopheles gambiae* s.s. *Insect Molecular Biology* 1998, 7(2):179-184.

98. Ranson H, Jensen B, Vulule JM, Wang X, Hemingway J, Collins FH: Identification of a point mutation in the voltage-gated sodium channel gene of Kenyan *Anopheles gambiae* associated with resistance to DDT and pyrethroids. *Insect Molecular Biology* 2000, 9(5):491-497.

99. O'Reilly AO, Khambay BP, Williamson MS, Field LM, Wallace BA, Davies TG: Modelling insecticide-binding sites in the voltage-gated sodium channel. *Biochem J* 2006, 396(2):255-263.

100. Jones CM, Liyanapathirana M, Agossa FR, Weetman D, Ranson H, Donnelly MJ, Wilding CS: Footprints of positive selection associated with a mutation (N1575Y) in the voltage-gated sodium channel of *Anopheles gambiae*. *Proc Natl Acad Sci U S A* 2012, 109(17):6614-6619.

101. Plapp FW, Jr.: The genetic basis of insecticide resistance in the house fly: evidence that a single locus plays a major role in metabolic resistance to insecticides. *Pesticide Biochemistry and Physiology* 1984, 22(2):194-201.

102. Georghiou GP, Ariaratnam V, Pasternak ME, Lin CS: Organophosphorus multiresistance in *Culex pipiens quinquefasciatus* in California. *Journal of Economic Entomology* 1975, 68(4):461-467.

103. Vontas J, David JP, Nikou D, Hemingway J, Christophides GK, Louis C, Ranson H: Transcriptional analysis of insecticide resistance in *Anopheles stephensi* using cross-species microarray hybridization. *Insect Mol Biol* 2007, 16(3):315-324.

104. Awolola TS, Oduola OA, Strode C, Koekemoer LL, Brooke B, Ranson H: Evidence of multiple pyrethroid resistance mechanisms in the malaria vector *Anopheles gambiae* sensu stricto from Nigeria. *Trans R Soc Trop Med Hyg* 2009, 103(11):1139-1145.

105. Wood O, Hanrahan S, Coetzee M, Koekemoer L, Brooke B: Cuticle thickening associated with pyrethroid resistance in the major malaria vector*Anopheles funestus*. *Parasit Vectors* 2010, 3:67.

106. Roberts DR, Chareonviriyaphap T, Harlan HH, Hshieh P: Methods of testing and analyzing excito-repellency responses of malaria vectors to insecticides. *J Am Mosq Control Assoc* 1997, 13(1):13-17.

107. Chandre F, Darriet F, Duchon S, Finot L, Manguin S, Carnevale P, Guillet P: Modifications of pyrethroid effects associated with *kdr* mutation in*Anopheles gambiae*. *Medical and Veterinary Entomology* 2000, 14(1):81-88.

108. Gahan JB, Lindquist AW: DDT residual sprays applied in buildings to control *Anopheles quadrimaculatus*. *Journal of Economic Entomology* 1945, 38 (2):223-230.

109. Chareonviriyaphap T, Roberts DR, Andre RG, Harlan HJ, Manguin S, Bangs MJ: Pesticide avoidance behavior in *Anopheles albimanus*, a malaria vector in the Americas. *J Am Mosq Control Assoc* 1997, 13(2):171-183.

110. Garros C, Marchand RP, Quang NT, Hai NS, Manguin S: First record of *Anopheles minimus* C and significant decrease of *An. minimus* A in central Vietnam. *J Am Mosq Control Assoc* 2005, 21(2):139-143.

111. Russell TL, Govella NJ, Azizi S, Drakeley CJ, Kachur SP, Killeen GF: Increased proportions of outdoor feeding among residual malaria vector populations following increased use of insecticide-treated nets in rural Tanzania. *Malar J* 2011, 10:80.

112. Moiroux N, Gomez MB, Pennetier C, Elanga E, Djenontin A, Chandre F, Djegbe I, Guis H, Corbel V: Changes in *Anopheles funestus* Biting Behavior Following Universal Coverage of Long-Lasting Insecticidal Nets in Benin. *J Infect Dis* 2012, 206(10):1622-1629.

113. WHO: Guidelines for testing mosquito adulticides intended for Indoor Residual Spraying (IRS) and Insecticide Treated Nets (ITNs). 2006, WHO/CDS/NTD/WHOPES/GCDDP/2006.3.

114. WHO: Report of the WHO Informal Consultation Tests procedures for insecticide resistance monitoring in malaria vectors, bio-efficacy and persistence of insecticides on treated surfaces. In. Geneva: World Health Organization: Parasitic Diseases and Vector Control (PVC)/Communicable Disease Control, Prevention and Eradication (CPE); 1998: 43.

115. Williams J, Pinto J: Training Manual on Malaria Entomology; For Entomology and Vector Control Technicians (Basic Level) In. Edited by USAID. Washington, D.C.; 2012: 78.

116. Ranson H, Abdallah H, Badolo A, Guelbeogo WM, Kerah-Hinzoumbe C, Yangalbe-Kalnone E, Sagnon N, Simard F, Coetzee M: Insecticide resistance in *Anopheles gambiae*: data from the first year of a multi-country study highlight the extent of the problem. *Malar J* 2009, 8(1):299.

117. Skovmand O, Bonnet J, Pigeon O, Corbel V: Median knock-down time as a new method for evaluating insecticide-treated textiles for mosquito control. *Malar J* 2008, 7:114.

118. Brogdon WG, McAllister JC: Simplification of adult mosquito bioassays through use of time-mortality determinations in glass bottles. *J Am Mosq Control Assoc* 1998, 14(2):159-164.

119. World Health Organization: Techniques to detect insecticide resistance mechanisms (field and laboratory manual). In. Edited by WHO/CDS/CPC/MAL/98.6 WHO. Geneva: World Health Organization; 1998.

120. Munhenga G, Masendu HT, Brooke BD, Hunt RH, Koekemoer LK: Pyrethroid resistance in the major malaria vector *Anopheles arabiensis* from Gwave, a malaria-endemic area in Zimbabwe. *Malar J* 2008, 7:247.

121. Okoye PN, Brooke BD, Koekemoer LL, Hunt RH, Coetzee M: Characterisation of DDT, pyrethroid and carbamate resistance in *Anopheles funestus*from Obuasi, Ghana. *Trans R Soc Trop Med Hyg* 2008, 102(6):591-598.

122. Kelly-Hope L, Ranson H, Hemingway J: Lessons from the past: managing insecticide resistance in malaria control and eradication programmes.*Lancet Infect Dis* 2008.

123. Chouaibou M, Etang J, Brevault T, Nwane P, Hinzoumbe CK, Mimpfoundi R, Simard F: Dynamics of insecticide resistance in the malaria vector*Anopheles gambiae* s.l. from an area of extensive cotton cultivation in Northern Cameroon. *Trop Med Int Health* 2008, 13(4):476-486.

124. Djegbe I, Boussari O, Sidick A, Martin T, Ranson H, Chandre F, Akogbeto M, Corbel V: Dynamics of insecticide resistance in malaria vectors in Benin: first evidence of the presence of L1014S *kdr* mutation in *Anopheles gambiae* from West Africa. *Malaria Journal* 2011, 10(1):261.

125. Dabire KR, Diabate A, Pare-Toe L, Rouamba J, Ouari A, Fontenille D, Baldet T: Year to year and seasonal variations in vector bionomics and malaria transmission in a humid savannah village in west Burkina Faso. *J Vector Ecol* 2008, 33(1):70-75.

126. Bass C, Nikou D, Donnelly MJ, Williamson MS, Ranson H, Ball A, Vontas J, Field LM: Detection of knockdown resistance (*kdr*)

mutations in*Anopheles gambiae*: a comparison of two new high-throughput assays with existing methods. *Malar J* 2007, 6:111.

127. Corbel V, N'Guessan R, Brengues C, Chandre F, Djogbenou L, Martin T, Akogbeto M, Hougard JM, Rowland M: Multiple insecticide resistance mechanisms in *Anopheles gambiae* and *Culex quinquefasciatus* from Benin, West Africa. *Acta Trop* 2007, 101(3):207-216.
128. Diabate A: The Role of Agricultural Uses of Insecticides in Resistance to Pyrethroids in *Anopheles gambiae* S.L. in Burkina Faso. *Am J Trop Med Hyg* 2002, 67(6):617-622.
129. Carnevale P, Toto JC, Guibert P, Keita M, Manguin S: Entomological survey and report of a knockdown resistance mutation in the malaria vector*Anopheles gambiae* from the Republic of Guinea. *Trans R Soc Trop Med Hyg*, 104(7):484-489.
130. Yawson AE, McCall PJ, Wilson MD, Donnelly MJ: Species abundance and insecticide resistance of *Anopheles gambiae* in selected areas of Ghana and Burkina Faso. *Med Vet Entomol* 2004, 18(4):372-377.
131. C. Fanello VP, A. della Torre, F. Santolamazza, G. Dolo, M. Coulibaly, A. Alloueche, C. F. Curtis, Y. T. Touré and M. Coluzzi: The pyrethroid knock-down resistance gene in the *Anopheles gambiae* complex in Mali and further indication of incipient speciation within *An. gambiae* s.s. *Insect Molecular Biology* 2003, 12(3):241-245.
132. Czeher C, Labbo R, Arzika I, Duchemin J-B: Evidence of increasing Leu-Phe knockdown resistance mutation in *Anopheles gambiae* from Niger following a nationwide long-lasting insecticide-treated nets implementation. *Malaria Journal* 2008, 7(1):189.
133. Awolola TS, Brooke BD, Hunt RH, Coetze M: Resistance of the malaria vector *Anopheles gambiae* s.s. to pyrethroid insecticides, in south-western Nigeria. *Annals of Tropical Medicine and Parasitology* 2002, 96(8):849-852.
134. Koffi AA, Alou LP, Adja MA, Kone M, Chandre F, N'Guessan R: Update on resistance status of *Anopheles gambiae* s.s. to conventional insecticides at a previous WHOPES field site, "Yaokoffikro", 6 years after the political crisis in Cote d'Ivoire. *Parasit Vectors* 2012, 5:68.

135. Dabire KR, Diabate A, Agostinho F, Alves F, Manga L, Faye O, Baldet T: Distribution of the members of *Anopheles gambiae* and pyrethroid knock-down resistance gene (*kdr*) in Guinea-Bissau, West Africa. *Bull Soc Pathol Exot* 2008, 101(2):119-123.

136. Etang J, Fondjo E, Chandre F, Morlais I, Brengues C, Nwane P, Chouaibou M, Ndjemai H, Simard F: First report of knockdown mutations in the malaria vector *Anopheles gambiae* from Cameroon. *Am J Trop Med Hyg* 2006, 74(5):795-797.

137. Ndjemai HN, Patchoke S, Atangana J, Etang J, Simard F, Bilong CF, Reimer L, Cornel A, Lanzaro GC, Fondjo E: The distribution of insecticide resistance in *Anopheles gambiae* s.l. populations from Cameroon: an update. *Trans R Soc Trop Med Hyg* 2009.

138. Nwane P, Etang J, Chouaibou M, Toto JC, Kerah-Hinzoumbe C, Mimpfoundi R, Awono-Ambene HP, Simard F: Trends in DDT and pyrethroid resistance in *Anopheles gambiae* s.s. populations from urban and agro-industrial settings in southern Cameroon. *BMC Infect Dis* 2009, 9:163.

139. Kerah-Hinzoumbe C, Peka M, Nwane P, Donan-Gouni I, Etang J, Same-Ekobo A, Simard F: Insecticide resistance in *Anopheles gambiae* from south-western Chad, Central Africa. *Malar J* 2008, 7:192.

140. Janeira F, Vicente JL, Kanganje Y, Moreno M, Do Rosario VE, Cravo P, Pinto J: A primer-introduced restriction analysis-polymerase chain reaction method to detect knockdown resistance mutations in *Anopheles gambiae*. *J Med Entomol* 2008, 45(2):237-241.

141. Mourou JR, Coffinet T, Jarjaval F, Pradines B, Amalvict R, Rogier C, Kombila M, Pages F: Malaria transmission and insecticide resistance of*Anopheles gambiae* in Libreville and Port-Gentil, Gabon. *Malar J* 2010, 9:321.

142. Himeidan YE, Chen H, Chandre F, Donnelly MJ, Yan G: Short report: permethrin and DDT resistance in the malaria vector *Anopheles arabiensis*from eastern Sudan. *Am J Trop Med Hyg* 2007, 77(6):1066-1068.

143. Abdalla H, Matambo TS, Koekemoer LL, Mnzava AP, Hunt RH, Coetzee M: Insecticide susceptibility and vector status of natural populations of*Anopheles arabiensis* from Sudan. *Transactions of the Royal Society of Tropical Medicine and Hygiene* 2008, 102(3):263-271.

144. Costantini C, Ayala D, Guelbeogo WM, Pombi M, Some CY, Bassole IH, Ose K, Fotsing JM, Sagnon N, Fontenille D *et al*: Living at the edge: biogeographic patterns of habitat segregation conform to speciation by niche expansion in *Anopheles gambiae*. *BMC Ecol* 2009, 9(1):16.

145. Kulkarni MA, Malima R, Mosha FW, Msangi S, Mrema E, Kabula B, Lawrence B, Kinung'hi S, Swilla J, Kisinza W *et al*: Efficacy of pyrethroid-treated nets against malaria vectors and nuisance-biting mosquitoes in Tanzania in areas with long-term insecticide-treated net use. *Trop Med Int Health* 2007, 12(9):1061-1073.

146. Kabula B, Tungu P, Matowo J, Kitau J, Mweya C, Emidi B, Masue D, Sindato C, Malima R, Minja J *et al*: Susceptibility status of malaria vectors to insecticides commonly used for malaria control in Tanzania. *Trop Med Int Health* 2012, 17(6):742-750.

147. Coleman M, Casimiro S, Hemingway J, Sharp B: Operational impact of DDT reintroduction for malaria control on *Anopheles arabiensis* in Mozambique. *J Med Entomol* 2008, 45(5):885-890.

148. Ratovonjato J, Le Goff G, Rajaonarivelo E, Rakotondraibe EM, Robert V: [Recent observations on the sensitivity to pyrethroids and DDT of*Anopheles arabiensis* and *Anopheles funestus* in the central Highlands of Madagascar; preliminary results on the absence of the *kdr* mutation in*An. arabiensis*]. *Arch Inst Pasteur Madagascar* 2003, 69(1-2):63-69.

149. Ramphul U, Boase T, Bass C, Okedi LM, Donnelly MJ, Muller P: Insecticide resistance and its association with target-site mutations in natural populations of *Anopheles gambiae* from eastern Uganda. *Trans R Soc Trop Med Hyg* 2009.

150. Verhaeghen K, Bortel WV, Roelants P, Okello PE, Talisuna A, Coosemans M: Spatio-temporal patterns in *kdr* frequency in permethrin and DDT resistant *Anopheles gambiae* s.s. from Uganda. *Am J Trop Med Hyg* 2010, 82(4):566-573.

151. Abate A, Hadis M: Susceptibility of *Anopheles gambiae* s.l. to DDT, malathion, permethrin and deltamethrin in Ethiopia. *Trop Med Int Health*2011, 16(4):486-491.

152. Ochomo E, Bayoh MN, Brogdon WG, Gimnig JE, Ouma C, Vulule JM, Walker ED: Pyrethroid resistance in *Anopheles gambiae* s.s. and *Anopheles arabiensis* in western Kenya: phenotypic,

metabolic and target site characterizations of three populations. *Med Vet Entomol* 2012.

153. Mathias DK, Ochomo E, Atieli F, Ombok M, Bayoh MN, Olang G, Muhia D, Kamau L, Vulule JM, Hamel MJ *et al*: Spatial and temporal variation in the *kdr* allele L1014S in *Anopheles gambiae* s.s. and phenotypic variability in susceptibility to insecticides in Western Kenya. *Malar J* 2011, 10:10.
154. Chanda E, Hemingway J, Kleinschmidt I, Rehman AM, Ramdeen V, Phiri FN, Coetzer S, Mthembu D, Shinondo CJ, Chizema-Kawesha E *et al*: Insecticide resistance and the future of malaria control in Zambia. *PLoS ONE* 2011, 6(9):e24336.
155. Mouatcho JC, Munhenga G, Hargreaves K, Brooke BD, Coetzee M, Koekemoer LL: Pyrethroid resistance in a major African malaria vector*Anopheles arabiensis* from Mamfene, northern KwaZulu-Natal, South Africa. *South African Journal of Science* 2009, 105(3-4):127-131.
156. Mouatcho JC, Hargreaves K, Koekemoer LL, Brooke BD, Oliver SV, Hunt RH, Coetzee M: Indoor collections of the *Anopheles funestus* group (Diptera: Culicidae) in sprayed houses in northern KwaZulu-Natal, South Africa. *Malar J* 2007, 6:30.
157. Brooke BD, Kloke G, Hunt RH, Koekemoer LL, Temu EA, Taylor ME, Small G, Hemingway J, Coetzee M: Bioassay and biochemical analyses of insecticide resistance in southern African *Anopheles funestus* (Diptera: Culicidae). *Bulletin of Entomological Research* 2001, 91(4):265-272.
158. Cuamba N, Morgan JC, Irving H, Steven A, Wondji CS: High level of pyrethroid resistance in an *Anopheles funestus* population of the Chokwe District in Mozambique. *PLoS ONE* 2010, 5(6):e11010.
159. Kloke RG, Nhamahanga E, Hunt RH, Coetzee M: Vectorial status and insecticide resistance of *Anopheles funestus* from a sugar estate in southern Mozambique. *Parasit Vectors* 2011, 4:16.
160. Hunt R, Edwardes M, Coetzee M: Pyrethroid resistance in southern African *Anopheles funestus* extends to Likoma Island in Lake Malawi. *Parasit Vectors* 2010, 3:122.
161. Wondji CS, Coleman M, Kleinschmidt I, Mzilahowa T, Irving H, Ndula M, Rehman A, Morgan J, Barnes KG, Hemingway J: Impact of pyrethroid resistance on operational malaria control in

Malawi. *Proc Natl Acad Sci U S A* 2012, 109(47):19063-19070.

162. Anto F, Asoala V, Anyorigiya T, Oduro A, Adjuik M, Owusu-Agyei S, Dery D, Bimi L, Hodgson A: Insecticide resistance profiles for malaria vectors in the Kassena-Nankana district of Ghana. *Malaria Journal* 2009, 8(1):81.
163. Djouaka R, Irving H, Tukur Z, Wondji CS: Exploring mechanisms of multiple insecticide resistance in a population of the malaria vector *Anopheles funestus* in Benin. *PLoS ONE* 2011, 6(11):e27760.
164. Dabire KR, Baldet T, Diabate A, Dia I, Costantini C, Cohuet A, Guiguemde TR, Fontenille D: *Anopheles funestus* (Diptera: Culicidae) in a humid savannah area of western Burkina Faso: bionomics, insecticide resistance status, and role in malaria transmission. *J Med Entomol* 2007, 44(6):990-997.
165. Faraj C, Adlaoui E, Brengues C, Fontenille D, Lyagoubi M: [Resistance of *Anopheles labranchiae* to DDT in Morocco: identification of the mechanisms and choice of replacement insecticide]. *Eastern Mediterranean health journal = La revue de sante de la Mediterranee orientale = al-Majallah al-sihhiyah li-sharq al-mutawassit* 2008, 14(4):776-783.
166. Mostafa AA, Allam KA: Studies on the present status of insecticides resistance on mosquitoes using the diagnostic dosages in El-Fayium Governorate, a spot area of malaria in Egypt. *J Egypt Soc Parasitol* 2001, 31(1):177-186.
167. Balkew M, Elhassan I, Ibrahim M, GebreMichael T, Engers H: Very high DDT-resistant population of *Anopheles pharoensis* Theobald (Diptera: Culicidae) from Gorgora, northern Ethiopia. *Parasite* 2006, 13(4):327-239.
168. Yadouleton AW, Padonou G, Asidi A, Moiroux N, Bio-Banganna S, Corbel V, N'Guessan R, Gbenou D, Yacoubou I, Gazard K *et al*: Insecticide resistance status in *Anopheles gambiae* in southern Benin. *Malar J* 2010, 9:83.
169. Pinto J, Lynd A, Vicente JL, Santolamazza F, Randle NP, Gentile G, Moreno M, Simard F, Charlwood JD, do Rosario VE *et al*: Multiple Origins of Knockdown Resistance Mutations in the Afrotropical Mosquito Vector *Anopheles gambiae*. *PLoS ONE* 2007, 2(11):e1243.

170. della Torre A, Fanello C, Akogbeto M, Dossou-yovo J, Favia G, Petrarca V, Coluzzi M: Molecular evidence of incipient speciation within *Anopheles gambiae* s.s. in West Africa. *Insect Mol Biol* 2001, 10(1):9-18.

171. Weill M, Chandre F, Brengues C, Manguin S, Akogbeto M, Pasteur N, Guillet P, Raymond M: The *kdr* mutation occurs in the Mopti form of*Anopheles gambiae* s.s. through introgression. *Insect Molecular Biology* 2000, 9(5):451-455.

172. Reimer LJ, Tripet F, Slotman M, Spielman A, Fondjo E, Lanzaro GC: An unusual distribution of the *kdr* gene among populations of *Anopheles gambiae* on the island of Bioko, Equatorial Guinea. *Insect Mol Biol* 2005, 14(6):683-688.

173. Protopopoff N, Verhaeghen K, Van Bortel W, Roelants P, Marcotty T, Baza D, D'Alessandro U, Coosemans M: A significant increase in *kdr* in*Anopheles gambiae* is associated with an intensive vector control intervention in Burundi highlands. *Trop Med Int Health* 2008, 13(12):1479-1487.

174. Pinto J, Lynd A, Elissa N, Donnelly MJ, Costa C, Gentile G, Caccone A, do Rosario VE: Co-occurrence of East and West African *kdr* mutations suggests high levels of resistance to pyrethroid insecticides in *Anopheles gambiae* from Libreville, Gabon. *Med Vet Entomol* 2006, 20(1):27-32.

175. Moreno M, Vicente JL, Cano J, Berzosa PJ, de Lucio A, Nzambo S, Bobuakasi L, Buatiche JN, Ondo M, Micha F *et al*: Knockdown resistance mutations (*kdr*) and insecticide susceptibility to DDT and pyrethroids in *Anopheles gambiae* from Equatorial Guinea. *Trop Med Int Health* 2008, 13(3):430-433.

176. Verhaeghen K, Van Bortel W, Roelants P, Backeljau T, Coosemans M: Detection of the East and West African *kdr* mutation in *Anopheles gambiae*and *Anopheles arabiensis* from Uganda using a new assay based on FRET/Melt Curve analysis. *Malaria Journal* 2006, 5(1):16.

177. Koekemoer LL, Spillings BL, Christian RN, Lo TC, Kaiser ML, Norton RA, Oliver SV, Choi KS, Brooke BD, Hunt RH *et al*: Multiple insecticide resistance in *Anopheles gambiae* (Diptera: Culicidae) from Pointe Noire, Republic of the Congo. *Vector Borne Zoonotic Dis* 2011, 11(8):1193-1200.

178. Reimer L, Fondjo E, Patchoke S, Diallo B, Lee Y, Ng A, Ndjemai HM, Atangana J, Traore SF, Lanzaro G *et al*: Relationship between *kdr* mutation and resistance to pyrethroid and DDT insecticides in natural populations of *Anopheles gambiae*. *J Med Entomol* 2008, 45(2):260-266.

179. Badolo A, Traore A, Jones CM, Sanou A, Flood L, Guelbeogo WM, Ranson H, Sagnon N: Three years of insecticide resistance monitoring in*Anopheles gambiae* in Burkina Faso: resistance on the rise? *Malar J* 2012, 11:232.

180. Santolamazza F, Calzetta M, Etang J, Barrese E, Dia I, Caccone A, Donnelly MJ, Petrarca V, Simard F, Pinto J *et al*: Distribution of knock-down resistance mutations in *Anopheles gambiae* molecular forms in west and west-central Africa. *Malar J* 2008, 7(1):74.

181. Ridl F, Bass C, Torrez M, Govender D, Ramdeen V, Yellot L, Edu A, Schwabe C, Mohloai P, Maharaj R *et al*: A pre-intervention study of malaria vector abundance in Rio Muni, Equatorial Guinea: Their role in malaria transmission and the incidence of insecticide resistance alleles. *Malaria Journal* 2008, 7(1):194.

182. Reimer L, Fondjo E, Patchok, Salomon, Diallo B, Lee Y, Ng A, Ndjemai HM, Atangana J, Traore SF *et al*: Relationship Between *kdr* Mutation and Resistance to Pyrethroid and DDT Insecticides in Natural Populations of *Anopheles gambiae*. *Journal of Medical Entomology* 2008, 45:260-266.

183. Matambo TS, Abdalla H, Brooke BD, Koekemoer LL, Mnzava A, Hunt RH, Coetzee M: Insecticide resistance in the malarial mosquito *Anopheles arabiensis* and association with the *kdr* mutation. *Medical and Veterinary Entomology* 2007, 21(1):97-102.

184. Kulkarni M, Rowland M, Alifrangis M, Mosha F, Matowo J, Malima R, Peter J, Kweka E, Lyimo I, Magesa S *et al*: Occurrence of the leucine-to-phenylalanine knockdown resistance (*kdr*) mutation in *Anopheles arabiensis* populations in Tanzania, detected by a simplified high-throughput SSOP-ELISA method. *Malaria Journal* 2006, 5(1):56.

185. Chen H, Githeko AK, Githure JI, Mutunga J, Zhou G, Yan G: Monooxygenase Levels and Knockdown Resistance (*kdr*) Allele

Frequencies in*Anopheles gambiae* and *Anopheles arabiensis* in Kenya. *Journal of Medical Entomology* 2008, 45:242-250.

186. Etang J, Manga L, Toto JC, Guillet P, Fondjo E, Chandre F: Spectrum of metabolic-based resistance to DDT and pyrethroids in *Anopheles gambiae*s.l. populations from Cameroon. *J Vector Ecol* 2007, 32(1):123-133.
187. Hargreaves K, Hunt RH, Brooke BD, Mthembu J, Weeto MM, Awolola TS, Coetzee M: *Anopheles arabiensis* and *An. quadriannulatus* resistance to DDT in South Africa. *Med Vet Entomol* 2003, 17(4):417-422.
188. Amenya DA, Naguran R, Lo TC, Ranson H, Spillings BL, Wood OR, Brooke BD, Coetzee M, Koekemoer LL: Over expression of a cytochrome P450 (CYP6P9) in a major African malaria vector, *Anopheles funestus*, resistant to pyrethroids. *Insect Mol Biol* 2008, 17(1):19-25.
189. Somboon P, Prapanthadara LA, Suwonkerd W: Insecticide susceptibility tests of *Anopheles minimus* s.l., *Aedes aegypti, Aedes albopictus*, and*Culex quinquefasciatus* in northern Thailand. *Southeast Asian J Trop Med Public Health* 2003, 34(1):87-93.
190. Verhaeghen K, Van Bortel W, Trung HD, Sochantha T, Coosemans M: Absence of knockdown resistance suggests metabolic resistance in the main malaria vectors of the Mekong region. *Malar J* 2009, 8:84.
191. Chareonviriyaphap T, Rongnoparut P, Chantarumporn P, Bangs MJ: Biochemical detection of pyrethroid resistance mechanisms in *Anopheles minimus* in Thailand. *J Vector Ecol* 2003, 28(1):108-116.
192. Rodpradit P, Boonsuepsakul S, Chareonviriyaphap T, Bangs MJ, Rongnoparut P: Cytochrome P450 genes: molecular cloning and overexpression in a pyrethroid-resistant strain of *Anopheles minimus* mosquito. *J Am Mosq Control Assoc* 2005, 21(1):71-79.
193. Verhaeghen K, Van Bortel W, Trung HD, Sochantha T, Keokenchanh K, Coosemans M: Knockdown resistance in *Anopheles vagus, An. sinensis, An. paraliae* and *An. peditaeniatus* populations of the Mekong region. *Parasit Vectors* 2011, 3(1):59.

194. Kang S, Jung J, Lee S, Hwang H, Kim W: The polymorphism and the geographical distribution of the knockdown resistance (*kdr*) of *Anopheles sinensis* in the Republic of Korea. *Malar J* 2012, 11:151.

195. Tan WL, Wang ZM, Li CX, Chu HL, Xu Y, Dong YD, Wang ZC, Chen DY, Liu H, Liu DP *et al*: First report on co-occurrence knockdown resistance mutations and susceptibility to beta-cypermethrin in *Anopheles sinensis* from Jiangsu Province, China. *PLoS ONE* 2012, 7(1):e29242.

196. Syafruddin D, Hidayati AP, Asih PB, Hawley WA, Sukowati S, Lobo NF: Detection of 1014F *kdr* mutation in four major Anopheline malaria vectors in Indonesia. *Malar J* 2010, 9:315.

197. Singh OP, Dykes CL, Das MK, Pradhan S, Bhatt RM, Agrawal OP, Adak T: Presence of two alternative *kdr*-like mutations, L1014F and L1014S, and a novel mutation, V1010L, in the voltage gated Na+ channel of *Anopheles culicifacies* from Orissa, India. *Malar J* 2010, 9:146.

198. Mishra AK, Chand SK, Barik TK, Dua VK, Raghavendra K: Insecticide resistance status in *Anopheles culicifacies* in Madhya Pradesh, central India.*J Vector Borne Dis* 2012, 49(1):39-41.

199. Sharma SN, Shukla RP, Raghavendra K: Susceptibility status of *An. fluviatilis* and *An. culicifacies* to DDT, deltamethrin and lambdacyhalothrin in District Nainital, Uttar Pradesh. *Indian J Malariol* 1999, 36(3-4):90-93.

200. Singh OP, Dykes CL, Lather M, Agrawal OP, Adak T: Knockdown resistance (*kdr*)-like mutations in the voltage-gated sodium channel of a malaria vector *Anopheles stephensi* and PCR assays for their detection. *Malar J* 2011, 10:59.

201. Tikar SN, Mendki MJ, Sharma AK, Sukumaran D, Veer V, Prakash S, Parashar BD: Resistance status of the malaria vector mosquitoes, *Anopheles stephensi* and *Anopheles subpictus* towards adulticides and larvicides in arid and semi-arid areas of India. *J Insect Sci* 2011, 11:85.

202. Baruah K, Lal S: A report on the susceptibility status of *Anopheles minimus* (Theobald) against DDT and deltamethrin in three districts of Assam. *J Vector Borne Dis* 2004, 41(1-2):42-44.

203. Karunaratne SH, Hemingway J: Malathion resistance and prevalence of the malathion carboxylesterase mechanism in populations of mosquito vectors of disease in Sri Lanka. *Bull World Health Organ* 2001, 79(11):1060-1064.

204. Kelly-Hope LA, Yapabandara AM, Wickramasinghe MB, Perera MD, Karunaratne SH, Fernando WP, Abeyasinghe RR, Siyambalagoda RR, Herath PR, Galappaththy GN *et al*: Spatiotemporal distribution of insecticide resistance in *Anopheles culicifacies* and *Anopheles subpictus* in Sri Lanka.*Trans R Soc Trop Med Hyg* 2005, 99(10):751-761.

205. Perera MD, Hemingway J, Karunaratne SP: Multiple insecticide resistance mechanisms involving metabolic changes and insensitive target sites selected in anopheline vectors of malaria in Sri Lanka. *Malar J* 2008, 7:168.

206. Surendran SN, Jude PJ, Weerarathne TC, Parakrama Karunaratne SH, Ramasamy R: Variations in susceptibility to common insecticides and resistance mechanisms among morphologically identified sibling species of the malaria vector *Anopheles subpictus* in Sri Lanka. *Parasit Vectors*2012, 5:34.

207. Mittal PK, Wijeyaratne P, Pandey S: Status of Insecticide Resistance of Malaria, Kala-azar and Japanese Encephalitis Vectors in Bangladesh, Bhutan, India and Nepal (BBIN). In. Edited by Project EH. Washington 2004.

208. Rowland M: Location of the gene for malathion resistance in *Anopheles stephensi* (Diptera: Culicidae) from Pakistan. *J Med Entomol* 1985, 22(4):373-380.

209. Abai MR, Mehravaran A, Vatandoost H, Oshaghi MA, Javadian E, Mashayekhi M, Mosleminia A, Piyazak N, Edallat H, Mohtarami F *et al*: Comparative performance of imagicides on *Anopheles stephensi*, main malaria vector in a malarious area, southern Iran. *J Vector Borne Dis* 2008, 45(4):307-312.

210. Lak SH, vatandoost H, Entezarmahdi MR, Ashraf H, Abai MR, Nazari M: Monitoring of Insecticide Resistance in *Anopheles sacharovi* (Favre, 1903) in Borderline of Iran, Armenia, Naxcivan and Turkey, 2001. *Iranian J Publ Health* 2002, 31(3-4):96-99.

211. Vatandoost H, Mashayekhi M, Abaie MR, Aflatoonian MR, Hanafi-Bojd AA, Sharifi I: Monitoring of insecticides resistance in main malaria vectors in a malarious area of Kahnooj district,

Kerman province, southeastern Iran. *J Vector Borne Dis* 2005, 42(3):100-108.

212. Malcolm CA: Current status of pyrethroid resistance in anophelines. *Parasitol Today* 1988, 4(7):S13-15.

213. Quinones ML, Suarez MF: Irritability to DDT of natural populations of the primary malaria vectors in Colombia. *J Am Mosq Control Assoc* 1989, 5(1):56-59.

214. Suarez MF, Quinones ML, Palacios JD, Carrillo A: First record of DDT resistance in *Anopheles darlingi*. *J Am Mosq Control Assoc* 1990, 6(1):72-74.

215. Fonseca-Gonzalez I: Estatus de la resistencia a insecticidas de los vectores primarios de malaria y dengue en Antioquia, Chocó, Norte de Santander y Putumayo, Colombia. Universidad de Antioquia, Colombia; 2008.

216. Fonseca-Gonzalez I, Cardenas R, Quinones ML, McAllister J, Brogdon WG: Pyrethroid and organophosphates resistance in *Anopheles (N.) nuneztovari* Gabaldon populations from malaria endemic areas in Colombia. *Parasitol Res* 2009, 105(5):1399-1409.

217. Chareonviriyaphap T, Golenda CF, Roberts DR, Andre RG: Identification of Elevated Esterase Activity in a Pyrethroid-Resistant Population of*Anopheles albimanus* Wiedemann. *ScienceAsia* 1999, 25 153-156.

218. Brogdon WG, McAllister JC, Corwin AM, Cordon Rosales C: Independent selection of multiple mechanisms for pyrethroid resistance in Guatemalan*Anopheles albimanus* (Diptera: Culicidae). *Journal of Economic Entomology* 1999, 92(2):298-302.

219. Zamora Perea E, Balta Leon R, Palomino Salcedo M, Brogdon WG, Devine GJ: Adaptation and evaluation of the bottle assay for monitoring insecticide resistance in disease vector mosquitoes in the Peruvian Amazon. *Malar J* 2009, 8:208.

220. Hemingway J, Penilla RP, Rodriguez AD, James BM, Edge W, Rogers H, Rodriguez MH: Resistance management strategies in malaria vector mosquito control. A large-scale field trial in Southern Mexico. *Pesticide Science* 1997, 51(3):375-382.

221. Dzul FA, Patricia Penilla R, Rodriguez AD: [Susceptibility and insecticide resistance mechanisms in *Anopheles albimanus* from the southern Yucatan Peninsula, Mexico]. *Salud Publica Mex* 2007, 49(4):302-311.

222. South Africa Department of Health: Malaria Updates. In. Pretoria, S.A: S.A.D.H.; 2003.

223. Maharaj R, Mthembu DJ, Sharp BL: Impact of DDT re-introduction on malaria transmission in KwaZulu-Natal. *S Afr Med J* 2005, 95(11):871-874.

224. Roberts DR, Manguin S, Mouchet J: DDT house spraying and re-emerging malaria. *Lancet* 2000, 356(9226):330-332.

225. Protopopoff N, Van Bortel W, Marcotty T, Van Herp M, Maes P, Baza D, D'Alessandro U, Coosemans M: Spatial targeted vector control in the highlands of Burundi and its impact on malaria transmission. *Malar J* 2007, 6:158.

226. Trape J-F, Tall A, Diagne N, Ndiath O, Ly AB, Faye J, Dieye-Ba F, Roucher C, Bouganali C, Badiane A *et al*: Malaria morbidity and pyrethroid resistance after the introduction of insecticide-treated bednets and artemisinin-based combination therapies: a longitudinal study. *The Lancet Infectious Diseases* 2011.

227. Lengeler C: Insecticide-treated bed nets and curtains for preventing malaria. *Cochrane Database of Systematic reviews* 2009(2):1-58.

228. Kitau J, Oxborough RM, Tungu PK, Matowo J, Malima RC, Magesa SM, Bruce J, Mosha FW, Rowland MW: Species shifts in the *Anopheles gambiae* complex: do LLINs successfully control *Anopheles arabiensis*? *PLoS ONE* 2012, 7(3):e31481.

229. Bradley J, Matias A, Schwabe C, Vargas D, Monti F, Nseng G, Kleinschmidt I: Increased risks of malaria due to limited residual life of insecticide and outdoor biting versus protection by combined use of nets and indoor residual spraying on Bioko Island, Equatorial Guinea. *Malar J* 2012, 11:242.

230. Henry MC, Assi SB, Rogier C, Dossou-Yovo J, Chandre F, Guillet P, Carnevale P: Protective efficacy of lambda-cyhalothrin treated nets in*Anopheles gambiae* pyrethroid resistance areas of Cote d'Ivoire. *Am J Trop Med Hyg* 2005, 73(5):859-864.

231. Corbel V, Akogbeto M, Damien GB, Djenontin A, Chandre F, Rogier C, Moiroux N, Chabi J, Banganna B, Padonou GG *et al*: Combination of malaria vector control interventions in pyrethroid resistance area in Benin: a cluster randomised controlled trial. *Lancet Infect Dis* 2012, 12(8):617-626.

232. Darriet F, N' Guessan R, Koffi AA, Konan L, Doannio JMC, Chandre F, Carnevale P: Impact of the resistance to pyrethroids on the efficacy of impregnated bednets used as a means of prevention against malaria: results of the evaluation carried out with deltamethrin SC in experimental huts. *Bulletin de la Société de Pathologie Exotique* 2000, 93(2):131-134.

233. Corbel V, Chandre F, Brengues C, Akogbeto M, Lardeux F, Hougard JM, Guillet P: Dosage-dependent effects of permethrin-treated nets on the behaviour of *Anopheles gambiae* and the selection of pyrethroid resistance. *Malar J* 2004, 3(1):22.

234. N'Guessan R, Corbel V, Akogbeto M, Rowland M: Reduced efficacy of insecticide-treated nets and indoor residual spraying for malaria control in pyrethroid resistance area, Benin. *Emerg Infect Dis* 2007, 13(2):199-206.

235. N'Guessan R, Asidi A, Boko P, Odjo A, Akogbeto M, Pigeon O, Rowland M: An experimental hut evaluation of PermaNet(R) 3.0, a deltamethrin-piperonyl butoxide combination net, against pyrethroid-resistant *Anopheles gambiae* and *Culex quinquefasciatus* mosquitoes in southern Benin.*Trans R Soc Trop Med Hyg* 2010, 104(12):758-765.

236. Asidi A, N'Guessan R, Akogbeto M, Curtis C, Rowland M: Loss of household protection from use of insecticide-treated nets against pyrethroid-resistant mosquitoes, Benin. *Emerg Infect Dis* 2012, 18(7):1101-1106.

237. Osse R, Gnanguenon V, Sezonlin M, Aikpon R, Padonou G, Yadouleton A, Akogbeto M: Relationship between the presence of *kdr* and Ace-1 mutations and the infection with *Plasmodium falciparum* in *Anopheles gambiae* s.s. in Benin. *Journal of Parassitology & Vector Biology* 2012, 4(3):31-39.

238. Moiroux N, Boussari O, Djenontin A, Damien G, Cottrell G, Henry MC, Guis H, Corbel V: Dry season determinants of malaria disease and net use in Benin, West Africa. *PLoS ONE* 2012, 7(1):e30558.

239. Poupardin R, Reynaud S, Strode C, Ranson H, Vontas J, David JP: Cross-induction of detoxification genes by environmental xenobiotics and insecticides in the mosquito *Aedes aegypti*: impact on larval tolerance to chemical insecticides. *Insect Biochem Mol Biol* 2008, 38(5):540-551.

240. IRAC: Prevention and Management of Insecticide Resistance in Vectors of Public Health Importance In: *Resistance Management for Sustainable Agriculture and Improved Public Health : Second Edition 2010* Insecticide Resistance Action Commitee; 2010: 72pp.

241. Georghiou GP, Taylor CE: Genetic and biological influences in the evolution of insecticide resistance. *Journal of Economic Entomology* 1977, 70(3):319-323.

242. Denholm I, Rowland MW: Tactics for managing pesticide resistance in arthropods: theory and practice. *Annu Rev Entomol* 1992, 37:91-112.

243. Djogbenou L, Weill M, Hougard JM, Raymond M, Akogbeto M, Chandre F: Characterization of insensitive acetylcholinesterase (ace-1R) in*Anopheles gambiae* (Diptera: Culicidae): resistance levels and dominance. *J Med Entomol* 2007, 44(5):805-810.

244. Berticat C, Bonnet J, Duchon S, Agnew P, Weill M, Corbel V: Costs and benefits of multiple resistance to insecticides for *Culex quinquefasciatus*mosquitoes. *BMC Evol Biol* 2008, 8:104.

245. Moore JH, Williams SM: Traversing the conceptual divide between biological and statistical epistasis: systems biology and a more modern synthesis. *Bioessays* 2005, 27(6):637-646.

246. Shono T, Zhang L, Scott JG: Indoxacarb resistance in the house fly, *Musca domestica*. *Pesticide Biochemistry and Physiology* 2004, 80(2):106-112.

247. Shono T, Kasai S, Kamiya E, Kono Y, Scott JG: Genetics and mechanisms of permethrin resistance in the YPER strain of house fly. *Pesticide Biochemistry and Physiology* 2002, 73(1):27-36.

248. Scott JG, Shono T, Georghiou GP: Genetic analysis of permethrin resistance in the house fly, *Musca domestica* L. *Experientia* 1984, 40(12):1416-1418.

249. Hardstone MC, Leichter CA, Scott JG: Multiplicative interaction between the two major mechanisms of permethrin resistance,

kdr and cytochrome P450-monooxygenase detoxification, in mosquitoes. *J Evol Biol* 2009, 22(2):416-423.

250. Berticat C, Boquien G, Raymond M, Chevillon C: Insecticide resistance genes induce a mating competition cost in *Culex pipiens* mosquitoes. *Genet Res* 2002, 79(1):41-47.
251. Agnew P, Berticat C, Bedhomme S, Sidobre C, Michalakis Y: Parasitism increases and decreases the costs of insecticide resistance in mosquitoes.*Evolution Int J Org Evolution* 2004, 58(3):579-586.
252. Foster SP, Harrington R, Devonshire AL, Denholm I, Devine GJ, Kenward MG: Comparative survival of insecticide-susceptible and resistant peach-potato aphids, *Myzus persicae* (Sulzer) (Hemiptera: Aphididae), in low temperature field trials. *Bull Ent Res* 1996, 86:17-27.
253. Shi MA, Lougarre A, Alies C, Fremaux I, Tang ZH, Stojan J, Fournier D: Acetylcholinesterase alterations reveal the fitness cost of mutations conferring insecticide resistance. *BMC Evol Biol* 2004, 4:5.
254. Djogbenou L, Noel V, Agnew P: Costs of insensitive acetylcholinesterase insecticide resistance for the malaria vector *Anopheles gambiae*homozygous for the G119S mutation. *Malar J* 2010, 9(1):12.
255. Brogdon WG, McAllister JC: Insecticide resistance and vector control. *Emerg Infect Dis* 1998, 4(4):605-613.
256. Diabate A, Baldet T, Chandre F, Akoobeto M, Guiguemde TR, Darriet F, Brengues C, Guillet P, Hemingway J, Small GJ *et al*: The role of agricultural use of insecticides in resistance to pyrethroids in *Anopheles gambiae* s.l. in Burkina Faso. *Am J Trop Med Hyg* 2002, 67(6):617-622.
257. Yadouleton A, Martin T, Padonou G, Chandre F, Asidi A, Djogbenou L, Dabire R, Aikpon R, Boko M, Glitho I *et al*: Cotton pest management practices and the selection of pyrethroid resistance in *Anopheles gambiae* population in northern Benin. *Parasit Vectors* 2011, 4:60.
258. Harrison G: Mosquitoes, malaria and man: A history of hostilities since 1880.; 1978.

259. Read AF, Lynch PA, Thomas MB: How to make evolution-proof insecticides for malaria control. *PLoS Biol* 2009, 7(4):e1000058.

260. Roush RT, Hoy CW, Ferro DN, Tingey WM: Insecticide resistance in the Colorado potato beetle (Coleoptera: Chrysomelidae): influence of crop rotation and insecticide use. *Journal of Economic Entomology* 1990, 83(2):315-319.

261. Georghiou GP: Insecticide resistance and prospects for its management. *Residue Reviews* 1980, 76:131-145.

262. Tabashnik BE: Managing resistance with multiple pesticide tactics: theory, evidence, and recommendations. *J Econ Entomol* 1989, 82(5):1263-1269.

263. Chitnis N, Schapira A, Smith T, Steketee R: Comparing the effectiveness of malaria vector-control interventions through a mathematical model. *Am J Trop Med Hyg* 2010, 83(2):230-240.

264. Yakob L, Dunning R, Yan G: Indoor residual spray and insecticide-treated bednets for malaria control: theoretical synergisms and antagonisms. *J R Soc Interface* 2011, 8(59):799-806.

265. World Health Organization: Global strategic framework for integrated vector management. Geneva; 2004.

266. Curtis CF: Theoretical models of the use of insecticide mixtures for management of resistance. *Bull Ent Res* 1985, 75: 259-265.

267. Hougard JM, Poudiougo P, Guillet P, Back C, Akpoboua LK, Quillevere D: Criteria for the selection of larvicides by the Onchocerciasis Control Programme in west Africa. *Ann Trop Med Parasitol* 1993, 87(5):435-442.

268. WHO: Pesticides and their application for the control of vectors and pests of public health importance; Sixth edition. In. Edited by WHO/CDS/NTD/WHOPES/GCDPP/2006.1 WHO, Geneva; 2006: 1-125.

269. Corbel V, Chabi J, Dabire RK, Etang J, Nwane P, Pigeon O, Akogbeto M, Hougard JM: Field efficacy of a new mosaic long-lasting mosquito net (PermaNet 3.0) against pyrethroid-resistant malaria vectors: a multi centre study in Western and Central Africa. *Malar J* 2010, 9:113.

270. Killeen GF, Okumu FO, N'Guessan R, Coosemans M, Adeogun A, Awolola S, Etang J, Dabire RK, Corbel V: The importance of

considering community-level effects when selecting insecticidal malaria vector products. *Parasit Vectors* 2011, 4:160.

271. Mani GS: Evolution of resistance in the presence of two insecticides. *Genetics* 1985, 109(4):761-783.
272. Roush RT: Designing resistance management programs: how can you choose? *Pesticide Science* 1989, 26(4):423-441.
273. Hougard JM, Corbel V, N'Guessan R, Darriet F, Chandre F, Akogbeto M, Baldet T, Guillet P, Carnevale P, Traore-Lamizana M: Efficacy of mosquito nets treated with insecticide mixtures or mosaics against insecticide resistant *Anopheles gambiae* and *Culex quinquefasciatus* (Diptera: Culicidae) in Cote d'Ivoire. *Bull Entomol Res* 2003, 93(6):491-498.
274. Asidi AN, N'Guessan R, Koffi AA, Curtis CF, Hougard JM, Chandre F, Corbel V, Darriet F, Zaim M, Rowland MW: Experimental hut evaluation of bednets treated with an organophosphate (chlorpyrifos-methyl) or a pyrethroid (lambdacyhalothrin) alone and in combination against insecticide-resistant *Anopheles gambiae* and *Culex quinquefasciatus* mosquitoes. *Malar J* 2005, 4(1):25.
275. Ohashi K, Nakada K, Ishiwatari T, Miyaguchi J, Shono Y, Lucas JR, Mito N: Efficacy of pyriproxyfen-treated nets in sterilizing and shortening the longevity of *Anopheles gambiae* (Diptera: Culicidae). *J Med Entomol* 2012, 49(5):1052-1058.
276. Mosqueira B, Duchon S, Chandre F, Hougard JM, Carnevale P, Mas-Coma S: Efficacy of an insecticide paint against insecticide-susceptible and resistant mosquitoes - part 1: laboratory evaluation. *Malar J* 2010, 9:340.
277. Ngufor C, N'Guessan R, Boko P, Odjo A, Vigninou E, Asidi A, Akogbeto M, Rowland M: Combining indoor residual spraying with chlorfenapyr and long-lasting insecticidal bed nets for improved control of pyrethroid-resistant *Anopheles gambiae*: an experimental hut trial in Benin. *Malar J*2011, 10:343.
278. Kleinschmidt I, Schwabe C, Shiva M, Segura JL, Sima V, Mabunda SJ, Coleman M: Combining indoor residual spraying and insecticide-treated net interventions. *Am J Trop Med Hyg* 2009, 81(3):519-524.

279. Okumu FO, Moore SJ: Combining indoor residual spraying and insecticide-treated nets for malaria control in Africa: a review of possible outcomes and an outline of suggestions for the future. *Malar J* 2011, 10:208.

280. Brosseau L, Drame PM, Besnard P, Toto JC, Foumane V, Le Mire J, Mouchet F, Remoue F, Allan R, Fortes F *et al*: Human antibody response to*Anopheles saliva* for comparing the efficacy of three malaria vector control methods in Balombo, Angola. *PLoS ONE* 2012, 7(9):e44189.

281. Bill_&_Melinda_Gates_Fondation, Boston_Consulting_Group: Market Assessment for Public Health Pesticide Products. In.; 2007.

Citations

CHAPTER 1

Philippe Nwane, Josiane Etang, Mouhamadou Chouaibou, Jean Claude Toto,Clément Kerah-Hinzoumbé, Rémy Mimpfoundi, Herman Parfait Awono-Ambene, and Frédéric Simard, Trends in DDT and Parathyroid Resistance in *Anopheles Gambiae* s.s. Populations from Urban and Agro-Industrial Settings in Southern Cameroon, doi:10.1186/1471-2334-9-163.

CHAPTER 2

Yallappa Rajashekar, Anjanappa Raghavendra, and Nandagopal Bakthavatsalam, " AcetylcholinesteraseInhibition by Biofumigant (Coumaran) from Leaves of Lantana camara in Stored Grain and Household Insect Pests," BioMed Research International, vol. 2014, Article ID 187019, 6 pages, 2014. doi:10.1155/2014/187019.

CHAPTER 3

Harsimran Kaur Gill and Harsh Garg (2014). Pesticides: Environmental Impacts and Management Strategies, Pesticides - Toxic Aspects, Dr. Sonia Soloneski (Ed.), ISBN: 978-953-51-1217-4, InTech, DOI: 10.5772/57399.

CHAPTER 4

Damien A. Devault and Hélène Pascaline (2014). Efficiency of Pesticide Alternatives in Non-agricultural Areas, Pesticides - Toxic Aspects, Dr. Sonia Soloneski (Ed.), ISBN: 978-953-51-1217-4, InTech, DOI: 10.5772/57150.

CHAPTER 5

Kathleen M. Raley-Susman (2014). Like a Canary in the Coal Mine: Behavioral Change as an Early Warning Sign of Neurotoxicological Damage, Pesticides - Toxic Aspects, Dr. Sonia Soloneski (Ed.), ISBN: 978-953-51-1217-4, InTech, DOI: 10.5772/57170.

CHAPTER 6

Vincent Corbel and Raphael N'Guessan (2013). Distribution, Mechanisms, Impact and Management of Insecticide Resistance in Malaria Vectors: A Pragmatic Review, Anopheles mosquitoes - New insights into malaria vectors, Prof. Sylvie Manguin (Ed.), ISBN: 978-953-51-1188-7, InTech, DOI: 10.5772/56117.

Index